Eric Bayerschen

Single-crystal gradient plasticity with an accumulated plastic slip: Theory and applications

Schriftenreihe
Kontinuumsmechanik im Maschinenbau
Band 9

Karlsruher Institut für Technologie (KIT)
Institut für Technische Mechanik
Bereich Kontinuumsmechanik

Hrsg. Prof. Dr.-Ing. habil. Thomas Böhlke

Eine Übersicht aller bisher in dieser Schriftenreihe erschienenen Bände finden Sie am Ende des Buchs.

Single-crystal gradient plasticity with an accumulated plastic slip: Theory and applications

by
Eric Bayerschen

Scientific Publishing

Dissertation, Karlsruher Institut für Technologie (KIT)
Fakultät für Maschinenbau
Tag der mündlichen Prüfung: 25. Oktober 2016

Impressum

KIT Scientific Publishing

Karlsruher Institut für Technologie (KIT)
KIT Scientific Publishing
Straße am Forum 2
D-76131 Karlsruhe

KIT Scientific Publishing is a registered trademark of Karlsruhe Institute of Technology. Reprint using the book cover is not allowed.

www.ksp.kit.edu

Print on Demand 2017

ISSN 2192-693X
ISBN 978-3-7315-0606-5
DOI 10.5445/KSP/1000062103

Single-crystal gradient plasticity with an accumulated plastic slip: Theory and applications

Zur Erlangung des akademischen Grades

Doktor der Ingenieurwissenschaften

der Fakultät für Maschinenbau

Karlsruher Institut für Technologie (KIT)

genehmigte

Dissertation

von

Dipl.-Ing. Eric Bayerschen

Tag der mündlichen Prüfung:	25.10.2016
Hauptreferent:	Prof. Dr.-Ing. Thomas Böhlke
Korreferent:	Prof. Dr. B. Daya Reddy

Zusammenfassung

Bei Versuchen an kleinskaligen Proben, wie metallischen Mikrodrähten, werden Größeneffekte beobachtet. Kleinere Proben reagieren dabei mechanisch steifer als größere Proben. In der vorliegenden Arbeit wird eine numerisch effiziente Gradientenplastizitätstheorie mittels der Finite Elemente Methode implementiert. Dieses Kontinuumsmodell wird dann auf Größeneffekte für kubisch-flächenzentrierte Metalle angewendet. Weiterhin werden Ergebnisse des Gradientenplastizitätsmodells mit Ergebnissen aus diskreten Versetzungsdynamiksimulationen verglichen. Dadurch können notwendige Erweiterungen der Theorie identifiziert werden, um das Modellverhalten zu verbessern.
Die Einleitung beinhaltet einen Überblick der Grundlagen der versetzungsbasierten Plastizität und der Gradientenplastizitätsmodellierung. Ein Kristallplastizitätsmodell wird mikromorph formuliert und mittels der Finite Elemente Methode implementiert. Als zusätzlicher Freiheitsgrad wird die akkumulierte plastische Gleitung berücksichtigt. Die isotrope Verfestigung wird unter Verwendung des Voce-Ansatzes modelliert. Numerische Ergebnisse dieses Modells werden dann verwendet, um experimentelle Beobachtungen an bambusartigen Mikrodrähten aus Gold zu erklären. Dies wird mit einer numerischen Untersuchung des Einflusses der Kristallorientierung auf die plastischen Feldverteilungen für zwei verschiedene Kristallorientierungen kombiniert.
Die Theorie wird erweitert, um den Gradienten der mikormorphen Variable und Korngrenzfließen zu berücksichtigen. Dieses Modell wird dann auf die Größeneffekte in Zugversuchen an wenig kristallinen Mikrodrähten aus Kupfer angewendet. Der Einfluss zwei verschiedener

Kristallorientierungen wird berücksichtigt, um die gegensätzlichen Größeneffekte in Zug- und Torsionsversuchen an Mikrodrähten aus Gold zu erläutern.
Durch den Vergleich von Ergebnissen des Gradientenplastizitätsmodells mit Ergebnissen aus diskreten Versetzungsdynamiksimulationen wird eine Erweiterung der Theorie durch Korngrenzverfestigung motiviert. Damit kann die Evolution der plastischen Dehnungen nahe der Korngrenzen vergleichsweise gut modelliert werden. Verbleibende Abweichungen der Gradientenverteilungen motivieren die Berücksichtigung einer verallgemeinerten Gradientenenergie im Modell. Es wird gezeigt, dass die Gradienten besser modelliert werden können, wenn diese verallgemeinerte Energieformulierung mit einem geeigneten Exponenten verwendet wird.
Der Kristallorientierungseinfluss auf die Korngrenzplastizitätsmechanismen wird in der vorliegenden Arbeit nicht berücksichtigt. Dennoch wird ein Überblick über die Literatur solcher Implementierungen gegeben. Weiterhin werden die geometrischen Transmissionskriterien für plastisches Gleiten von Versetzungen beschrieben. Die Limitierungen des aktuellen Modells werden diskutiert und es werden Vorschläge zur Entwicklung einer physikalisch weiterentwickelten Gradientenplastizitätstheorie gemacht.

Summary

On small-scale specimen such as metallic microwires, size effects are experimentally observed with smaller specimens responding mechanically stronger, compared to larger specimens. In this work, a numerically efficient gradient plasticity theory is implemented using the finite element method. The continuum model is applied to size effects on face-centered cubic metals. In addition, gradient plasticity results are compared to discrete dislocation dynamics results. Thereby, extensions of the theory are identified in order to improve the model behavior.
The introduction is comprised of an overview on dislocation-based plasticity fundamentals and on gradient crystal-plasticity modeling. Subsequently, a micromorphic crystal-plasticity finite element implementation is presented, using an accumulated plastic slip as additional degree of freedom and a Voce-formulation for isotropic hardening. Numerical results of this framework are used to explain experimental observations on bamboo-structured gold microwires. This is combined with a numerical investigation of the crystal-orientation influence on the plastic field distributions for two distinct crystal orientations.
The theory is extended by consideration of the gradient of the micromorphic variable and by grain boundary yielding. This model is applied to the size effects observed in tensile test experiments on oligocrystalline copper microwires. The influence of two distinct crystal-orientations is considered and the contrary size effects experimentally observed on oligocrystalline gold microwires are, thereby, elucidated.
A comparison of gradient plasticity results to discrete dislocation dynamics results motivates the extension of the model by additional grain

boundary hardening. Thereby, the evolution of plastic strain near the grain boundaries is modeled comparably well. Remaining deviations in the gradient distributions motivate the consideration of a generalized defect energy in the model. It is shown that the gradients can be modeled better, using this generalized energy form with a suitable exponent. The orientation-dependence of grain boundary plasticity mechanisms is not considered in the theory. However, an overview is given on the literature on computational implementations of such concepts in conjunction with a critical review of geometrical transmission criteria for plastic slip of dislocations. The limitations of the current gradient plasticity model are addressed and suggestions are made for the development of a physically more advanced gradient plasticity theory.

Acknowledgments

I would like to express my gratitude to Prof. Thomas Böhlke for offering me a position as an academic employee, for advising me during the past years, for trusting in me, my teaching and research, and for his enduring support including stays abroad and conference participations.
Furthermore, I would like to thank Prof. Daya Reddy for co-advising and supporting this work, and for the offer to work for three months at the Centre for Research and Computational Mechanics (CERECAM) of the University of Cape Town in South Africa during the past doctorate.
I would also like to thank the members of the DFG research group FOR1650 for the fruitful discussions and the collaborative work. Especially, I thank Markus Stricker for all his work regarding discrete simulations and joint publications, Michael Ziemann for providing experimental data and images, and Daniel Weygand as well as Patric Gruber and Mario Walter for their comprehensive contributions to the joint publications. I would like to thank Andrew McBride for taking the time to discuss my work and for his support of the publication written during my stay at CERECAM. Former students Elvedin Ramani, (and now-colleagues) Andreas Prahs and Hannes Erdle are acknowledged for their work and effort towards collaborative publications and talks. I thank Moritz Wenk for providing a labor-intensive, experimentally obtained dislocation image for the introduction chapter.
Then, I would like to thank all colleagues and former colleagues at the Institute of Engineering Mechanics and at CERECAM who all have created a work environment that felt almost like being with family. First and foremost, I am grateful to my office colleagues Vedran Glavas,

Malte Schemmann, and Stephan Wulfinghoff for their support, "open ears", and for the welcomed distraction at times. I would like to especially thank Stephan Wulfinghoff for advising me during the early stages and for attentively explaining theoretical concepts as well as the art of programming in C++ to me. The model used and extended in this work is based on his implementation of a gradient plasticity theory, within a finite element code written and developed by Felix Fritzen. Therefore, I would like to thank Felix Fritzen very much for the opportunity to use his highly efficient code and for his assistance in many aspects of programming, implementation, postprocessing and other Linux-related troubles. I thank Katja Jöchen for her support during the early stages of my doctoral thesis. Barthel Brylka and Rudolf Neumann are acknowledged for providing their expertise in hands-on applied mechanics and (exhaust) fume physics. Next, I thank Ute Schlumberger-Maas and Helga Betsarkis for taking care of all the administrative effort regarding conference organization, travel expenses, paper submissions, proofreading, and many things more. I wish to thank Tom-Alexander Langhoff for providing the computational infrastructure and for all his guidance with regard to Ubuntu. He always had an "open ear", also for questions of teaching and engineering mechanics. Florian Rieger is acknowledged for proofreading several publications which expedited the publication process significantly.

Personally, I thank my parents Rita and Bruno, my sister Ramona, Liese, Yannik, Moritz, Jacqueline and Christoph, Nina and Johannes for their wholehearted encouragement and support.

The financial support by the German Research Foundation (DFG) under Grants BO1466/5-1/5-2, part of the DFG Research Group 1650 "Dislocation based Plasticity", and by the Karlsruhe House of Young Scientists (KHYS) for funding a stay abroad are gratefully acknowledged.

Karlsruhe, October 2016 Eric Bayerschen

Contents

Motivation

Metals are an important class of materials for a wide range of industrial applications due to their reliability and strength. Applications include, for example, sheet metals for automotive parts such as chassis or cast metals for engine blocks. In recent years, also more and more application of metals in miniaturized medical devices such as, for instance, cardiac pacemakers has gained much importance. In order to guarantee safety of use, durability, and customer-satisfying lifetime of the products with a minimum material effort, a key engineering-discipline has become the prediction of the behavior of materials, including metals, and of their deformation under loading.

Therefore, many material models have been developed during the 20th century. Determination of the material properties and required model parameters in experiments, however, is a cost and time-intensive process. Furthermore, during development, examination of products and components under different loading conditions is necessary and has traditionally been accomplished with experimental setups. Due to the increasing availability and performance of personal computers, however, simulation approaches have been expedited and are commonly used, nowadays. The benefits of such simulations are apparent: after the material model has been calibrated to a reference experiment, further simulations can be carried out, reducing the necessity of additional experiments. Consequently, a broad variety of loading conditions can easily be applied and tested just by means of simulations. The physical quantities of interest, e.g., critical stresses, necessary to identify potential spots of failure of the product, can be obtained using computations.

In this context, the investigation of the plastic behavior, i.e., the permanent deformation of materials is of large interest. This includes the development of efficient manufacturing processes such as deep-drawing and, for example, loading-dependent lifetime prediction of products. Defects in the atomic structure of the material govern the plastic deformation behavior. The movement of the defects and their interactions result in local stress-inhomogeneities that, in turn, influence the overall mechanical behavior of the materials. Metals (and many other materials) are composed of periodic arrangements of atoms that are ordered in unit cells, i.e., of crystals. It is possible to produce metallic parts such that all unit cells are uniquely oriented. The material is then of single-crystalline (lattice) structure. Usually, however, after the casting process, metals are composed of many regions with different crystal orientations, called grains. The interfaces between grains, the grain boundaries (GBs), substantially influence the material behavior by, e.g., trapping defects in their vicinity.
The fundamental carrier of plastic deformation are line-defects called dislocations. These move through the crystal lattice, attract and entangle or repel each other. They can pile up at GBs or at other interfaces such as precipitates. The discovery of dislocations, during the early 20th century, motivated researchers to consider these in their material models. Continuum approaches have been developed to model the dislocation-based plastic behavior. In such models, the dislocations are accounted for in an averaged sense, i.e., not individual dislocations but rather ensembles of dislocations are modeled. Single-crystal plasticity theories are an example, modeling the movement of dislocations by considering plastic slip on the different slip systems of the crystalline unit structure. The work-hardening of the material is predicted, for example, by a constitutive relation between the resolved shear stresses and a scalar quantity describing the plastic deformation. They are commonly used in today's industry and the scientific community. Discrete simulation approaches such as discrete dislocation dynamics have also

been developed, modeling in detail the reactions and interactions of the individual dislocations. These discrete models, however, are often computationally expensive due to the discretization of all individual dislocations present in the simulation volume. Thus, their use for large-scale simulations is limited or, at least, time-consuming.

Continuum models offer substantial benefit in terms of fast computational times due to the averaged representation of dislocations. The classic plasticity theories have been used successfully for many problems such as, e.g., for prediction of hardening behavior due to plastic anisotropy or for prediction of texture evolution. However, their use is limited when the plastic deformation of materials is localized and strongly inhomogeneous. For example, it has been observed that, for micron-sized wires, the grain size of the specimen influences the overall mechanical response. Such phenomena are not present for larger wires with, e.g., cross-section diameters of a few mm. While the classic plasticity theories can be used for larger specimen with an at least overall homogeneous plastic deformation, they fail to predict the mechanical behavior adequately in the micro-range. This is due to the influence arising from microstructural characteristics such as GBs. During plastic deformation, dislocations pile up at the GBs leading to inhomogeneous deformation stages and local stress concentrations preventing other dislocations to move further through the lattice. These inhomogeneities lead to an increase in the overall mechanical strength, e.g., the macro yield-stress, due to the arising higher local stresses required to advance the plastic deformation.

A particular class of continuum theories, single-crystal gradient plasticity theories, has been developed and pushed forward during the last four decades. In these theories, the non-locality of plastic deformation is considered by, e.g., taking into account the gradients of plastic slips. In contrast to classic crystal-plasticity, phenomena such as stiffer mechanical responses of microwires due to, e.g., decreasing grain sizes can be modeled. Due to their consideration of non-locality, these models possess an internal length scale that the classic plasticity approaches lack. However, within a full multi-slip scenario, e.g., for materials with face-centered cubic crystal structure, these theories become increasingly computationally expensive due to the necessity to evaluate all gradients of the plastic slips. Therefore, in some contributions in the literature, models have been presented based on a scalar measure of plastic slip, and its gradient. Thereby, the computational benefit of classic single-crystal theories can, for the most part, be maintained while still allowing for modeling inhomogeneous plastic deformation phenomena, to some extent. As a consequence, fully three-dimensional simulations are feasible, even for aggregates composed of several grains.

The point of departure for this thesis is a geometrically linear gradient plasticity theory using a single scalar quantity as an additional degree of freedom to describe the plastic deformation. It has been published in Wulfinghoff and Böhlke (2012), and been extended by a grain boundary yield condition in Wulfinghoff et al. (2013). A first calibration to experimental results looked promising (Wulfinghoff et al., 2013), although the texture of the specimens (i.e., the crystal orientation of the grains) could not be considered. However, several open questions regarding the model, its applicability, and its further development remained. These include:

- Can the use of one scalar quantity and its gradient to describe the plastic deformation give physically meaningful results, i.e., can experimental results be reproduced?
- Is this model feasible for the prediction of experimentally observed complex phenomena such as contrary size-effects under different loading conditions?
- What is the influence of the microstructural characteristics, such as the crystal orientations of the different grains, on these gradient-attributed phenomena?

- How does the continuum model perform in comparison to physically more advanced, discrete dislocation-based models?
- Which enhancements of the theory are necessary in order to model discrete dislocation mechanisms in an averaged manner on the continuum scale?

In the present thesis, a step is taken towards clarification of these questions, identification of the model applicability and its limitations, and an enhancement of the theory.

State of the art

Dislocations in metals

In this chapter, an overview is given on dislocation-based plasticity theories, focusing on metals. The state of the art of gradient crystal plasticity for modeling dislocation-based plasticity on the continuum scale is presented. Parts of this summary are taken from Bayerschen et al. (2015), Bayerschen and Böhlke (2016), and Bayerschen et al. (2016a).

From a historical point of view, the discovery of dislocations revolutionized the material modeling of metals. This started during the early 20th century with works by Orowan (1934a;b;c); Polanyi (1934); Taylor (1934) (see also Hirth, 1985, for a brief overview on the history of dislocation-related theories). In conjunction with theories of crystallographic shear, phenomena such as the difference between the theoretical and the measured shear strength (Schmid and Boas, 1935) or hardening (Taylor, 1938) could be explained. Burgers (1939) developed a vector field theory for the elastic dislocation fields. The concept of smearing discrete dislocations into a continuous array of infinitesimal dislocation goes back to a work by Brown Jr (1941), including the description of a net-density of dislocations. Later on, Nye (1953) developed a mathematical connection between the net-dislocation density tensor and the lattice curvature of a crystal. Kröner (1959) introduced the concept of incompatibility. Anisotropic elastic theories of dislocations originate in the pioneering work by Eshelby et al. (1953). The idea of Frank-Read sources was developed by Frank and Read Jr (1950), and the existence

of dislocations could be proven using the in the 1930s developed experimental method of transmission electron microscopy (Heidenreich, 1949; Bollmann, 1956; Hirsch et al., 1956).
Important early works in the context of the presence of GBs and their influence on the dislocation movement, and on the resulting mechanical response, are the ones by Read and Shockley (1950); Bilby et al. (1964). It was discovered that the presence of GBs in microstructured materials can lead to "non-classic" plastic behavior with regard to the strength of the material (e.g., specimens with smaller grain sizes respond stiffer than specimens with larger grain sizes, Sylwestrowicz and Hall, 1951; Hall, 1951a;b; Petch, 1953). Furthermore, dislocation interactions could be shown to exist, also with GBs (see, e.g., Kacher et al., 2014, for an overview). A review of the literature on criteria used to predict the transmission of dislocations (and plastic slip) at GBs can be found in Bayerschen et al. (2016a).
Theories of hardening were developed including phenomenological approaches (e.g., Koehler, 1952; Feltham and Meakin, 1957; Kocks, 1976; Estrin and Mecking, 1984). These account, for example, for mobile dislocations becoming randomly trapped during their movement. Chaboche (1989) proposed a theory for kinematic hardening, in order to explain the cyclic mechanical response of metals. The phenomenological theories are handy in terms of their limited number of parameters which can be calibrated by experiments (e.g., Seeger et al., 1957; Kovács and Feltham, 1963; Mughrabi, 1978; Zehetbauer and Seumer, 1993). During the early stages of dislocation-based models for hardening, no distinction was made between contributions by dislocations stored randomly in the bulk and by dislocations accommodating inhomogeneous plastic deformations. Experimental and theoretical works from the 1970s on (Hirth, 1972; Thompson et al., 1973; Ronay, 1979), however, have shown that many observable effects can be attributed to the latter, geometrically necessary, dislocations. Their influence on the material response is non-negligible when the deformations are comparably inhomogeneous.

The experimental investigation of dislocation and grain-size related size effects is still an ongoing task with more recent works by Fleck et al. (1994); Chen and Ngan (2011); Yang et al. (2012); Chen et al. (2015).
A broad variety of modeling approaches for dislocations and the associated interactions and mechanical phenomena has emerged with the improvement of computational resources (see Dingreville et al., 2016, for a recent overview on computational modeling and experimental characterization across the length scales).

Atomistic and discrete dislocation modeling approaches

In atomistic approaches, dislocations are modeled based on atomic and interatomic potentials (e.g., Mishin et al., 1998; Lee et al., 2003). Therefore, the interactions of dislocations can be investigated in a detailed fashion also with regard to obstacles such as GBs (e.g., Bitzek and Gumbsch, 2005; Bachurin et al., 2010). Discrete dislocation dynamics (e.g., Weygand et al., 2002; 2009; Šiška et al., 2009) model the interactions of dislocations without resolving all atoms in the simulation volume. Instead, the dislocation line segments are discretized and interaction rules are used to model multiplication, annihilation and other interactions such as trapping of dislocations at interfaces. Due to the physically detailed modeling, atomistics and discrete dislocation dynamics can be used to "feed" multi-scale models (e.g., Dewald and Curtin, 2007a), or to benchmark continuum models (e.g., Stricker et al., 2016). The physically detailed modeling, however, often comes with the drawback of increased computational times, prohibiting the direct use for large-scale applications.

Continuum modeling of dislocations

Continuum models do not resolve dislocations individually. Instead, the mechanical behavior of ensembles of dislocations is modeled, often in a phenomenological way. Examples of macroscopic material models include, for example, the theories by Hill (1966); Teodosiu and Sidoroff (1976); Asaro (1983); Needleman and Tvergaard (1993). Such theories are implemented, for example, with finite elements and simulations are carried out to support experimental findings (e.g., Yao et al., 2014; Ziemann et al., 2015; Guery et al., 2016). Early computational works considered only a reduced number of slip systems and date back to the 1980s (e.g., Peirce et al., 1982, see Roters (2011) for a more detailed treatise of the historic perspective). Crystal-plasticity models offer comparably low computational time requirements and a broad spectrum of applicability to single- as well as polycrystalline materials. Applications include, for example, the prediction of hardening behavior due to plastic anisotropy (Beyerlein et al., 2007) and texture evolution (Eyckens et al., 2015), modeling phenomena such as deformation twinning (Kalidindi, 1998), and shear banding (Forest, 1998). Whenever the material response is of single-crystalline type (Zaafarani et al., 2006) or at least polycrystalline such that individual grains are not predominant regarding the effective material response (Zhang et al., 2015), macroscopic continuum approaches are valuable to use since the microstructure of the material does not need to be accounted for, explicitly. In cases when microstructural characteristics of the material become predominant, continuum models need to be enriched by additional considerations, such as the incorporation of strain gradients (Tian et al., 2014), and by explicit modeling of the influence resulting from interfaces, e.g., in bimetallic materials (Mayeur et al., 2015).

The classic plasticity theories, however, fail to model phenomena such as size effects due to their lack of an internal length scale (Hutchinson, 2000). Therefore, the interest of the community in the modeling of these effects has been much increased and resulted in the emergence of a variety of gradient plasticity theories.

Gradient plasticity theories for single crystals

Some of the first works using gradient plasticity approaches to describe the mechanical behavior of single crystals or crystal-ensembles were published by Aifantis (1984; 1987). The gradient-extended crystal plasticity theories can be classified into work-conjugate and non-work-conjugate theories (Kuroda and Tvergaard, 2006). In the work-conjugate theories, the free energy is extended by an additional contribution. This so-called defect- or gradient energy considers, for example, the gradients of plastic slips or plastic strain. The need to consider this contribution arises from the coarsening error made in the continuum formulation of the discrete dislocation ensembles (Mesarovic, 2010). In the non-work-conjugate theories, however, additional hardening stresses are posed on the individual slip systems based on physical considerations of the collective dislocation behavior. It can be shown that both types of approaches are equivalent within a three-dimensional multislip context (Kuroda and Tvergaard, 2008). Therefore, both types of theories can be used to predict the same kind of length-scale related effects in the material response.

In gradient-extended continuum models, conservative glide of dislocations is often assumed and dislocation transfer across interfaces such as GBs is modeled, usually, from a phenomenological perspective (e.g., Aifantis et al., 2006). Thereby, the discrete causes of, for example, strain fields close to GBs can, however, not be distinguished anymore. For instance, continuum models incorporating GB yielding cannot distinguish between strain fields caused by dislocation transmission across the GB and strain fields caused by absorption of dislocations from adjacent grains into the GB (Zhang et al., 2014). However, more sophisticated continuum models have been developed that incorporate physical mechanisms such as, for example, climbing of dislocations (Geers et al., 2014). The model of van Beers et al. (2015a) considers the redistribution of defects along the GBs in an averaged sense via a diffusion-type equation for the spreading of the net-defect content of the GBs along their planar surfaces. Dislocation transport is considered in the models of Reuber et al. (2014); Dogge et al. (2015). In the latter work, flux equations are explicitly accounted for at the interfaces, modeling the transport across them. It has been experimentally determined that the changes in line energy, accompanying dislocation motion, are important in the context of GB dislocations (Lucadamo and Medlin, 2002). The framework by Wulfinghoff and Böhlke (2015) takes into account the transport of dislocations in the bulk and curvature-induced line-length production by coupling a physically enriched continuum dislocation dynamics model (Hochrainer et al., 2014) with a simplified gradient plasticity model (Wulfinghoff et al., 2013).

Modeling GB mechanisms within gradient plasticity theories is an ongoing challenge, e.g., Aifantis and Willis (2005); Fredriksson and Gudmundson (2005); Gurtin (2008); Van Beers et al. (2013). It is known that five macroscopic and three microscopic degrees of freedom are necessary to specify a general GB (Wolf, 1990). Continuum models, however, commonly neglect the microscopic degrees of freedom, i.e., the translations between slip systems from adjacent grains at a GB are not accounted for due to the coarsening made. Only the macroscopic degrees of freedom are then considered which define the rotations between slip systems and the GB, respectively. For large-grained microstructures (Kacher et al., 2014), the GB modeling is of utmost importance since the GB presence and influence on the dislocation movement leads to pile-ups of dislocations that can, in turn, dominate the material behavior.

Size effects can be modeled in gradient plasticity models by different approaches. The free energy can be enhanced by terms taking into account excess dislocations (Gurtin, 2000; 2002). Thereby, gradient stresses or back-stresses are induced which enter, for example, the equations for the plastic slip rates. Such approaches neglect the influence of GBs or other obstacles on the dislocation structures. If, for example, dislocation structures in dual-phase steels are to be described on the grain scale taking into account interaction of ferrite grains and the coverage of ferrite grains by martensite particles, then such an approach is insufficient (Rieger and Böhlke, 2015). The aforementioned grain interactions can be incorporated by a GB yield condition (Wulfinghoff et al., 2013) which mimics the slip interaction on GBs by additional constitutive equations which can take into account the misorientation between grains (Van Beers et al., 2013), the orientation of the GB relative to the grain orientations (Gurtin, 2008), and the GB defect structure and energy (van Beers et al., 2015b;c). By such GB yield mechanisms, a (grain) size effect is induced in the mechanical model response.

Nowadays, the computational resources allow more and more to implement material models with many degrees of freedom, e.g., Gottschalk et al. (2016). Numerical efficiency, however, is still of utmost importance in order to not loose the computational benefits of continuum models with regard to time consumption. A numerical treatment of a strain gradient plasticity theory has been presented in Niordson and Kysar (2014). The model is of viscous type, and considers both dissipative and energetic contributions. This is in contrast to frameworks, such as the one treated in Reddy et al. (2012), which consider purely energetic contributions. Within the implementation by Özdemir and Yalçinkaya (2014) of the gradient plasticity theory by Gurtin (2008), all plastic slips are considered as additional degrees of freedom and interface elements are used for the GB discretization. Results are presented for the two-dimensional case. In Gottschalk et al. (2016), however, an implementation for three dimensions is discussed and numerical examples are shown for both flow rules proposed by Gurtin (2008) as well as for an additionally proposed form. The implementation of

the theory is extended to finite deformations in McBride et al. (2016). Variational formulations for single-crystal gradient plasticity at large deformations are presented in Reddy (2013).
The approach Ettehad and Al-Rub (2015) suggests using a two-step solution scheme for gradient plasticity, separating the computations into a global solution and a local approximation that facilitates meshfree methods. It is discussed that this two-step procedure simplifies the implementation of gradient theories, with an emphasis on the enforcement of boundary conditions.
Recently, a variational framework with thermomechanical coupling for finite strains has been proposed by Bartels et al. (2015). The approach by Anand et al. (2015) accounts for thermal annealing, and a computational study of this model, including a viscous regularization, is presented in McBride et al. (2015). In the recent work by Lubarda (2016), the fraction of the rate of plastic work converted into heat is incorporated in the free energy of two gradient plasticity formulations to account for the "locked-in strain energy" around statistically stored dislocations. Both isothermal and non-isothermal settings and boundary conditions are explored, there.
In the spirit of keeping the numerical costs comparably low, a simplified gradient plasticity theory has been proposed by Wulfinghoff and Böhlke (2012); Wulfinghoff et al. (2013), considering only one additional degree of freedom to incorporate the plastic effects. Due to numerical reasons, a micromorphic approach, suggested by Forest (2009), is used for the implementation of this theory.

Micromorphic approaches

In micromorphic approaches, additional variables are introduced as additional internal degrees of freedom. Micromorphic theories belong to the class of generalized continuum theories. It can be formally shown that gradient theories, which also belong to this class, are special cases of micromorphic theories (see Forest and Sievert (2003), and also Lazar (2010) for an overview on dislocations in generalized continua frameworks). In Forest (2009), a general framework for the micromorphic approach is outlined. It includes balance equations governing the micromorphic degrees of freedom, boundary conditions, and higher-order stresses (see also the references in Forest, 2009). An overview of applications to, e.g., elasticity and gradient plasticity is also given. In this context, it is described in detail that models formulated with the micromorphic approach are related to existing gradient models. Therefore, the micromorphic variable is constrained to be equal to its (macro) counterpart (see also Forest, 2016). This constraint can be imposed by a penalty term in the free energy. The constrained micromorphic approach yields models that, e.g., belong to the class of gradient of internal variable models (Maugin, 1990). For instance, the gradient theory by Gurtin (2003) can be regarded as a constrained micromorphic theory (Forest, 2009). Micromorphic crystal-plasticity or gradient-plasticity theories have recently been evaluated in comparison to discrete dislocation dynamics (e.g., Bayerschen et al., 2015; Chang et al., 2016).

Defect energy in gradient plasticity theories

For the incorporation of the defect energy, several approaches exist. An overview of energy formulations is given in Forest (2016), for micromorphic theories. The shared aim of continuum theories considering a defect energy is the modeling of lattice distortion due to geometrically necessary dislocations (GNDs), for example, by consideration of the dislocation density tensor (Nye, 1953). It is, however, outlined in Mesarovic (2010) that the common association of the gradients to GNDs via the dislocation density tensor is not unique. In Gurtin (2000), a geometrically non-linear theory is proposed using a defect energy chosen "for convenience" to be quadratic in the plastic part of the deformation gradient. A quadratic defect energy is also used in Cermelli and Gurtin (2002), formulated to depend on the geometrical dislocation density tensor within a geometrically linear theory. This formulation reduces to a quadratic form in the Burgers vector for strict plane strain. The discrete dislocation dynamics results of Nicola et al. (2005) include a comparison of several energy formulations. Their results motivated Gurtin et al. (2007) to reformulate the defect energy with dependence on the densities of screw and edge dislocations rather than on the Burgers tensor. It is shown there, that this defect energy can be recast in terms of the gradients of plastic slips. The discussion of scaling regimes by Scardia et al. (2014) includes this defect energy by Gurtin et al. (2007) in their intermediate scaling regime.

A defect energy that is quadratic in the geometric dislocation density tensor has also been used in the recent variational implementations of finite gradient plasticity by Miehe (2014); Miehe et al. (2014b;a). In Miehe (2014), constitutive rate-type and algorithmic incremental potentials are defined in order to formulate a general framework of inelasticity, applicable, e.g., to construct single-crystal gradient plasticity. In line with this framework, a computational approach to gradient plasticity (of von Mises-type in the logarithmic strain space), using mixed variational principles, is proposed in Miehe et al. (2014b). This includes a separation into variables of long-range and short-range character, i.e., of macro-motion and micro-motion, respectively. In the variational approach by Miehe et al. (2014a), such a separation is performed for multiplicative plasticity and a viscous regularization technique is proposed to overcome the problems of classic active-set search for rate-independent plasticity. In addition, mixed variational principles have been exploited for small strains in the previous work Miehe et al. (2013).

Apart from the commonly used quadratic defect energy formulations, non-quadratic forms have also been proposed in the literature. In

Ohno et al. (2008), a defect energy is introduced that is linear in the accumulated GND-densities, thereby leading to a constant higher-order (micro) stress. A defect energy of more general type is used by Gurtin and Ohno (2011). In their work, several special cases are discussed where they distinguish between recoverable and nonrecoverable defect energies. Recoverable defect energies in this regard means that loading and reverse loading, starting from a set of slip gradients, leads to the same value of the defect energy. Nonrecoverable defect energy means, however, that the same value of the defect energy is not necessarily obtained after such a loading cycle. In Reddy (2011b), it is shown that the defect energy proposed by Ohno and Okumura (2007) (see also Kametani et al., 2012), is recoverable but non-differentiable for vanishing slip gradients. It is also shown there, that the defect energy by Ohno et al. (2008), formulated linear in the accumulated dislocation densities, is, however, recoverable. In Ohno and Okumura (2007), it is pointed out that the higher-order (micro) stress resulting from their defect energy has the form of a step function. This behavior is in contrast to the resulting behavior from the quadratic defect energies which lead to gradient stresses changing linearly in dependence of, e.g., the slip gradients. Furthermore, these two different defect energy approaches are interpreted physically in Ohno and Okumura (2007). The consideration of dislocation self-energy leads to a constant gradient stress. Consideration of the dislocation interaction-energy, however, leads to a gradient stress changing with slip gradients, and, thereby, accounting for the strain hardening induced by dislocation pile-ups. In Hurtado and Ortiz (2012), non-local effects are accounted for in the deformation of micropillars by considering the energy of dislocation surface steps, in addition to the self-energy of dislocations. The thermo-mechanical framework of Anand et al. (2015) also uses the linear defect energy by Ohno and Okumura (2007) in the mechanical contribution to the free energy. In Reddy (2011a), the influence of the defect energy and of internal-variable hardening is discussed with regard to the well-posedness of the mathematical problem statement for several theories from the literature.

Linear as well as quadratic defect energies are reviewed in Forest and Guéninchault (2013), and it is shown that the latter ones lead to physically realistic slip profiles but also to unusual scaling laws. Instead, a logarithmic defect energy is proposed and connected to the statistical theory of Groma et al. (2003). Such a logarithmic defect energy is further investigated in Wulfinghoff et al. (2015) and compared to a linear defect energy by analytical calculations. In addition, numerical calculations with a quadratic regularization of both energy approaches are performed to resolve the issues of differentiability for vanishing GND-densities. In the recent review by Forest (2016), the different energy forms are contrasted, also in regard to the obtained scaling laws. A non-convex strain gradient plasticity model for patterning is proposed in Yalcinkaya et al. (2011). Although a quadratic defect energy is used, an additional polynomial in terms of plastic slips is employed in the free energy. The choice of the value of the internal length scale is shown to affect the distributions of plastic slip. This model is extended in Yalçinkaya et al. (2012) using a non-convex latent hardening formulation by Ortiz and Repetto (1999).

A generalized power-law type defect energy is proposed by Bardella (2010) and Bardella et al. (2013). It is formulated in dependence of Nye's dislocation density tensor. This dependence has originally been proposed by Gurtin (2002). In Bayerschen and Böhlke (2016), a power-law type defect energy using an accumulated plastic slip and its gradient is investigated with regard to size effects of crystal aggregates. The work of Voyiadjis et al. (2014) considers short-range and long-range dislocation interactions in the defect energy of their theory. There, it is discussed, with reference to Bardella (2010), that the exponent of the proposed defect energy governs the non-linearity of the defect energy approach (see also Bayerschen and Böhlke, 2016). In the recent work of Bardella and Panteghini (2015), it is proposed to consider the defect energy as a function of two invariants of Nye's tensor. A power-law defect energy is proposed for this dependence and investigated for the choice of a quadratic defect energy exponent. It is also shown, there, that the experimental results by Fleck et al. (1994) can be fitted with a regularized logarithmic defect energy. Recent atomistic simulations (Begau et al., 2015) and theoretical works (Kooiman et al., 2015; 2016; Berdichevsky, 2016) present indications that the mathematical form of the defect energy could be of log-linear type (see also the discussion in Forest, 2016).

In other recent works, it is discussed that further ingredients (aside from the defect energy) should be considered in the development of appropriate crystal plasticity models accounting for length scale effects. These include, for example, dislocation transport (Reuber et al., 2014) and vacancy diffusion in addition to coupled dislocation glide-climb mechanisms (Geers et al., 2014).

Outline of the thesis

In Chapter 1, the fundamentals of dislocation-based continuum plasticity are summarized, including gradient-extended single-crystal plasticity modeling. Then, in Chapter 2, results obtained with a basic micromorphic crystal-plasticity model are compared to experimental quasi-single-crystalline microwire torsion-test results. Subsequently, in Chapter 3, the model is extended by consideration of gradient plasticity with grain boundary yielding, and size effects on oligocrystalline microwires are modeled. The theory is supplemented by grain boundary hardening and simulation results are compared to discrete dislocation dynamics simulation results in Chapter 4. In Chapter 5, the defect energy approach is generalized, and the implications on the mechanical model response are shown. Then, in Chapter 6, an overview on slip transmission criteria is given and the limitations of the model to consider these are addressed. A summary in Chapter 7, featuring also a brief discussion of remaining open questions, concludes this thesis.

Notation

A direct tensor notation is preferred throughout the text. Vectors and 2nd-order tensors are denoted by bold letters, e.g., by $\boldsymbol{a}$ or $\boldsymbol{A}$. A linear mapping of 2nd-order tensors by a 4th-order tensor is written as $\boldsymbol{A} = \mathbb{C}[\boldsymbol{B}]$. The scalar product and the dyadic product are denoted, e.g., by $\boldsymbol{A} \cdot \boldsymbol{B}$ and $\boldsymbol{A} \otimes \boldsymbol{B}$, respectively. The composition of two 2nd-order tensors is formulated with $\boldsymbol{AB}$. The 2nd-order unity tensor is denoted by $\boldsymbol{I}$. Matrices are denoted by a hat, e.g., by $\hat{\varepsilon}$. The full list of symbols and used operators can be found on pp. 224.
It is remarked that throughout this thesis the quantity γ_{ac}, the accumulated plastic slip, is used. In the journal publications corresponding to contents of several chapters of this thesis, this quantity is termed "equivalent plastic strain" and denoted by γ_{eq}.

Chapter 1

Introduction

1.1 Dislocation-induced plastic deformation

It is known from experimental works that, with increasing loading, plastically deformed material can require higher applied stresses to advance the plastic deformation further. This material behavior is called hardening, see Fig. 1.1, as opposed to softening.

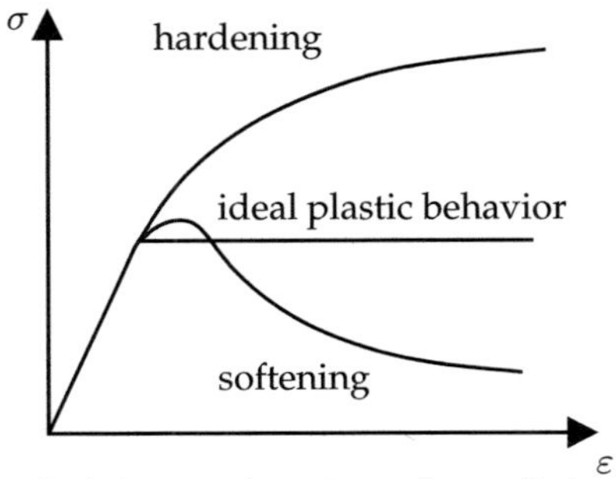

Figure 1.1: Nominal stress σ of specimens for applied nominal strain ε.

Both phenomena are caused by the presence, generation, movement, and interaction of defects in the atomic structure of the material. Defects of line-like character are called dislocations. These can be experimentally observed, see, for example, Fig. 1.2.

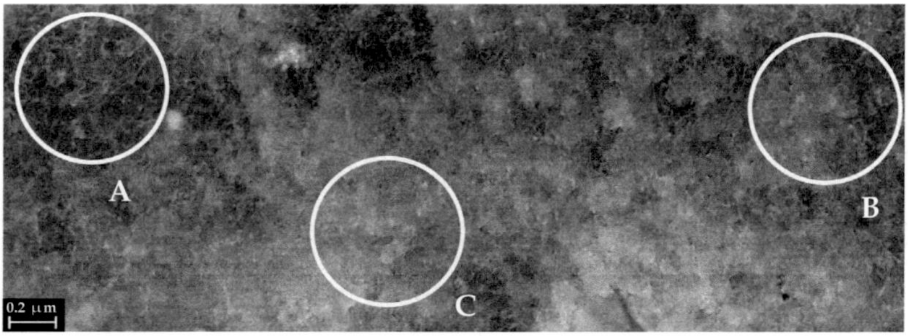

Figure 1.2: Visible dislocation structures in a ferrite grain of a dual-phase steel obtained by electron channeling contrast imaging. **A**, **B**: dislocation loops and segments. **C**: dislocation forest. Preparation and image courtesy of M. Wenk, Institute of Applied Materials, Karlsruhe Institute of Technology (KIT).

When the material is macroscopically plastically deformed, the deformation is locally resolved by plastic slip of the dislocations (Essmann et al., 1968). This phenomenon affects metals but also other crystalline materials, e.g., calcite (De Bresser, 1996). The crystal lattice is locally sheared in certain slip directions, depending on the orientation of the crystal with regard to the applied loading, and dislocation movement becomes observable in form of slip traces, see, e.g., Lim and Raj (1985a). From a theoretical point of view, a dislocation can be defined as in the following described procedure. An ideal, defect-free crystal lattice of atoms is schematically depicted in Fig. 1.3a. The atoms are ordered in equidistant positions. In contrast, the crystal lattice in the presence of a dislocation is depicted in Fig. 1.3b. Upon removal of several atoms from one of the lattice lines, i.e., formally, the introduction of a dislocation, there, elastic deformation occurs in the immediate vicinity of the dislocation until an equilibrium state of the atomic positions has been reached (Fig. 1.3b). This change in position affects atoms up to a few atomic layers around the dislocation. Further away from the dislocation, the crystal lattice is not affected anymore and retains the unaltered lattice structure (see also Fig. 1.3b). The dislocation can be characterized by its so-called Burgers vector. This can be obtained by evaluating the Burgers circuit which is constituted by a line integral around the dislocation in the elastically deformed state, see Fig. 1.3c. Starting at a chosen atom (highlighted in the figure in blue color), a closed line segment around the dislocation can be defined. However, if the elastic deformation is relaxed, i.e., the atoms assume their original, unaltered positions, the same line segment exhibits a gap, Fig. 1.3d. The vector associated to the closure-failure between the two ends of the line segment is called the Burgers vector. Its magnitude b characterizes the shortest distance between both ends. A vector $\tilde{\boldsymbol{l}}$ can be attached to the direction of the dislocation line, see Fig. 1.3c (in the present example, $\tilde{\boldsymbol{l}}$ is perpendicular to the drawing plane).

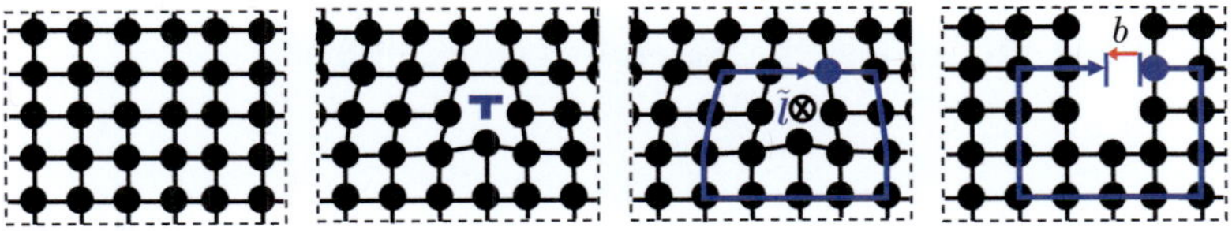

(a) (b) (c) (d)

Figure 1.3: (a) Undeformed crystal lattice. (b) Deformed crystal lattice in the presence of a dislocation. (c) Burgers circuit around a dislocation in the deformed crystal lattice. The line direction is indicated by $\tilde{\boldsymbol{l}}$. (d) Burgers circuit around a dislocation in the undeformed crystal lattice.

Mathematically, the definition of the Burgers vector can be given in terms of the following line integral (e.g., Shetty, 2013)

$$\boldsymbol{b} = \oint \frac{\mathrm{d}\boldsymbol{u}_\mathrm{e}}{\mathrm{d}\tilde{s}} \,\mathrm{d}\tilde{s}. \tag{1.1}$$

The vector $\boldsymbol{u}_\mathrm{e}$ denotes the (elastic) displacement, and $\tilde{s}$ is the variable of the circuit. According to this definition, the sign of the Burgers vector depends on the convention employed for the direction of the parametrization variable $\tilde{s}$ of the line integral, see Fig. 1.4. Dislocation lines cannot end within the bulk material (Hull and Bacon, 2011).

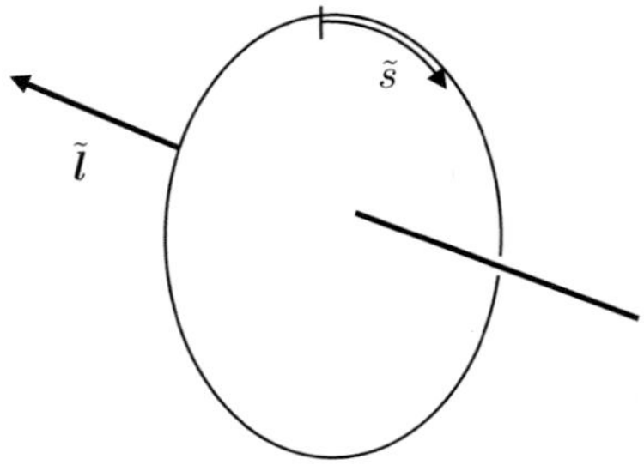

Figure 1.4: Burgers circuit, parametrized by $\tilde{s}$, around a dislocation line segment with line direction $\tilde{\boldsymbol{l}}$ in the deformed lattice.

They either form closed loops, inside the material, or end at the outside surface or at interfaces, also inside of the material. For junctions of dislocations, this implies the necessary condition that

$$\sum_{i=1}^{n} \boldsymbol{b}_i = 0, \tag{1.2}$$

where $\boldsymbol{b}_i$ are the Burgers vectors of the dislocations $i = 1, \ldots, n$.
Idealized, two different types of dislocations can be distinguished. Based on the orientation of the Burgers vector $\boldsymbol{b}$ with regard to the line direction $\tilde{\boldsymbol{l}}$ of the dislocation, the dislocation is either of edge-type or screw-type. For edge dislocations, $\boldsymbol{b} \perp \tilde{\boldsymbol{l}}$, and, for screw dislocations, $\boldsymbol{b} \parallel \tilde{\boldsymbol{l}}$. In general, however, dislocations are curved and can be composed of different line segments from both types.
During plastic deformation, dislocations in the crystal lattice move on specific planes, depending on the crystal structure. The smallest periodic entity is named unit cell. Many metals posses the face-centered cubic (FCC) unit cell structure (Fig. 1.5a). Examples include copper, nickel, aluminum, and gold. The planes are located in between closest-packed atomic layers. They are called slip planes and are defined by their normal vectors $\boldsymbol{n}$. The direction in which the dislocation moves on such a plane is called slip direction $\boldsymbol{d}$. Both vectors, $\boldsymbol{n}$ and $\boldsymbol{d}$, constitute a

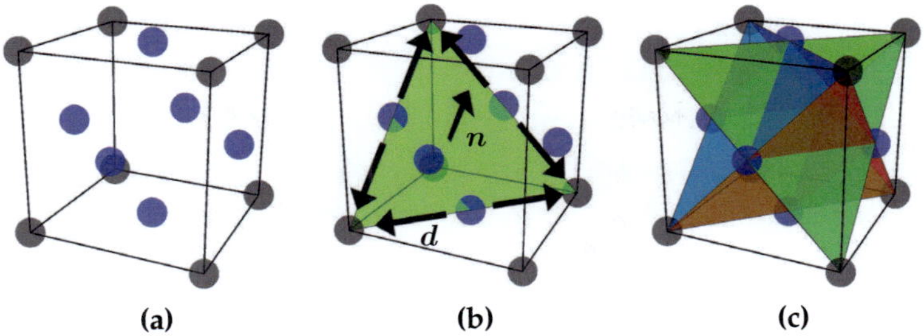

Figure 1.5: (a) Face-centered cubic unit cell. (b) Slip directions $\boldsymbol{d}$ and normal $\boldsymbol{n}$ of exemplary slip plane. (c) Slip planes of face-centered cubic unit cell. Basic visualization tool courtesy of V. Glavas.

slip system. An example of a slip plane for the FCC unit cell is depicted in Fig. 1.5b. The full set of FCC slip planes is shown in Fig. 1.5c. For this type of crystal unit cell, four planes of closest-packed atomic layers exist with three possible slip directions each. Taking into account that dislocations can move in both positive and negative direction with respect to $\boldsymbol{d}$, dislocations can move on 24 different combinations of slip planes and slip directions.

1.2 Interaction of dislocations

As the plastic deformation proceeds, more and more dislocations are generated by dislocation sources (e.g., by Frank-Read sources), and the dislocations progressively entangle due to the increasing density of dislocations. They also interact with other defects present in the atomic structure. These include voids, precipitates, substitutional and interstitial atoms. In addition, dislocations pile up, for example, at interfaces that they encounter during their movement. These pile-ups lead to an increase in the stresses necessary to move the dislocations through the lattice. Consequently, the interactions of dislocations can impede their motion. These interactions are complex processes.

For example, dislocations can be resolved into partial dislocations. They can also annihilate other dislocations of opposite character. At obstacles, dislocation climb is also a common mechanism. Depending on the loading and the microstructure, dislocations are generated, interact, and accumulate by random trapping in the bulk material. These dislocations are referred to as statistically stored dislocations (SSDs) (Ashby, 1970). At interfaces, however, dislocations distribute such that an excess of dislocations with identical sign can be observed. These dislocations are required for compatible deformation and are called geometrically necessary dislocations (GNDs). They occur due to gradients of the plastic shear in the material (Nye, 1953) that are caused by the geometry of the loading or by inhomogeneous deformations. Both types of dislocations influence the hardening behavior of the material.

1.3 Dislocation behavior at grain boundaries

The interfaces between different grains, i.e., between regions of different crystal orientations, are called grain boundaries (GBs). In general, GBs are curved but they also exhibit regions of planar type. Grain boundaries can impede the motion of dislocations in which case these tend to pile up, there. In the following, the basic dislocation mechanisms near GBs are briefly discussed. Parts of this section are taken from Bayerschen et al. (2016a).

In Fig. 1.6a, a schematic of a dislocation pile-up at a GB with normal $\boldsymbol{n}_\Gamma$ is depicted for a single slip system. Here, $\boldsymbol{n}^{\mathrm{A}}$ is the slip plane normal of the depicted slip system in grain A, and $\boldsymbol{d}^{\mathrm{A}}$ is a slip direction within the slip plane. The dislocations encounter resistance against their motion by GBs, due to the mismatch of the crystal lattices of the adjacent grains, there. This resistance is affected by both geometrical and physical influences. First of all, the above-mentioned mismatch in the crystal lattices at GBs leads to a mismatch of slip planes and directions, there.

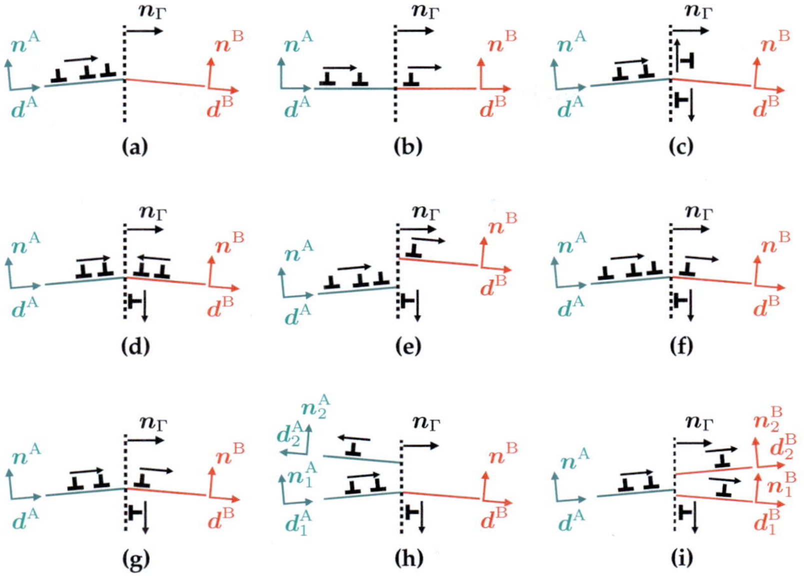

Figure 1.6: Schematic: (a) Pile-up of dislocations. (b) Ideal transmission of a dislocation. (c) Dissociation of a dislocation into the grain boundary. (d) Absorption of two dislocations into the grain boundary, generating one dislocation. (e) Re-emission of dissociated dislocation. (f) Generation and emission of a dislocation into grain B and generation of a residual dislocation. (g) Direct transmission of dislocation, and generation of a residual dislocation. (h) Reflection of a dislocation at the grain boundary and generation of a residual dislocation. (i) Dislocation splitting up onto different slip systems and generation of a residual dislocation. Figures (a-h) reprinted from Bayerschen et al. (2016a) with permission from Springer.

The resistance is also influenced by the orientation of the GBs, i.e., by their inclinations, and by the structure of the atomic layers close to the GBs, which differs from the ideal lattice. This is the case because of the elastic deformations due to the different atom positions in the adjacent grains. As a result, GBs pose barriers for the movement of dislocations. Furthermore, trapped dislocations at GBs can impose an additional barrier for the passing of dislocations. It has been found that, the better the alignment of slip systems from the adjacent grains at a GB, the higher is the probability of dislocations passing the GB,

or of activating dislocation sources on the adjacent slip systems. Ideal alignment of two slip systems and a direct transfer of a dislocation is schematically shown in Fig. 1.6b.
In real plastic deformation processes, many more rather complex mechanisms take place at GBs. For instance, dislocations from one grain (Fig. 1.6c) or from both grains (Fig. 1.6d) can be resolved into the GB. The deposited dislocations influence arriving dislocations or can be re-emitted (Fig. 1.6e). Grain boundaries can also act as a source for the generation of new dislocations that subsequently move onto the adjacent slip systems (Fig. 1.6f). If a dislocation is transmitted to a slip system of different orientation, a residual dislocation is generated in the GB (Fig. 1.6g). The reason for the generation of residual dislocations is the necessary continuity of the Burgers vector, Eq. (1.2). Depending on the orientation of the line integral around the dislocation, the residual Burgers vector (RBV) can be defined, e.g., by (Lim and Raj, 1985c)

$$\boldsymbol{b}_{\mathrm{r}} + \boldsymbol{b}^{\mathrm{B}} = \boldsymbol{b}^{\mathrm{A}}. \tag{1.3}$$

This residual Burgers vector remains in the GB and ensures the continuity of the Burgers vector across the GB. Dislocation mechanisms at GBs usually involve more than one slip system in each grain. Exemplarily, the reflection of a dislocation at a GB is depicted in Fig. 1.6h, and a dislocation splitting up onto different slip systems is shown in Fig. 1.6i. More involved mechanisms at GBs include, for example, the absorption of dislocations into the GBs by either dissociation of the dislocation into displacement-shift-complete dislocations (Pond and Smith, 1977), suitable for the specific type of GB (Clark and Smith, 1979), or by remaining as localized dislocations. Furthermore, which of these mechanisms become active (some may also act simultaneously Shen et al., 1988) depends not only on the crystal-lattice orientations of both grains and of the GB but also on the type of dislocations, i.e., whether these are of edge, screw, or mixed type. Screw dislocations can, in principle,

cross GBs without leaving a residual dislocation (Lim and Raj, 1985b). Continuous screw slip-bands across GBs are more likely to occur due to a transfer of dislocations than due to activation of dislocation sources in the adjacent grain (Lim and Raj, 1985a).
Transmission of dislocations across GBs is a transfer of line defects and can be associated to the transfer of plastic slip by means of the Orowan equation (Orowan, 1934a). In experiments, the activity of slip systems in the grains adjacent to a GB can be investigated by analysis of the slip traces (Lall et al., 1979) which show different degrees of continuity or discontinuity across GBs (Seal et al., 2012; West and Was, 2013). A discontinuous slip trace is schematically shown in Fig. 1.7a.

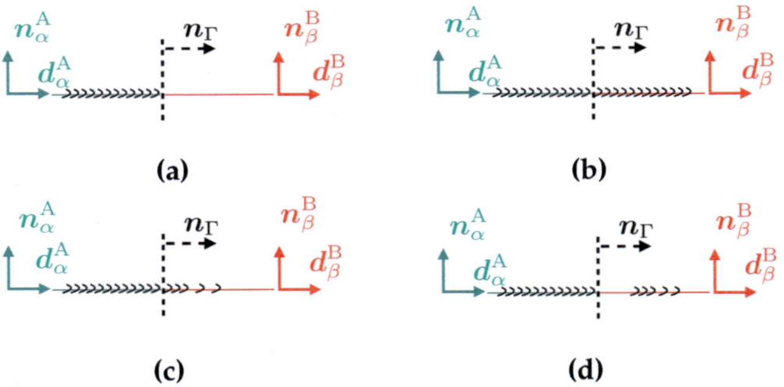

Figure 1.7: (a) Discontinuity and (b-d) continuity of slip traces and dislocation movement across a grain boundary with normal n_Γ between two slip systems α, β in grains A, B. Figure reprinted from Bayerschen et al. (2016a) with permission from Springer.

For instance, in West and Was (2013), some of the developing slip traces, in the grain adjacent to a grain with a dislocation pile-up, were observed to spread over the whole grain (see schematic in Fig. 1.7b), but others reached only a few microns into the adjacent grain (see schematic in Fig. 1.7c). The authors of the latter work propose to classify slip traces into the continuous and the discontinuous type, and to consider all GBs intersected by dislocation channels for this evaluation.

This consideration also includes GBs where dislocation channels stopped to develop further into the adjacent grain on the incoming grain-side of the GB but new channels developed a few microns away from the GB in the adjacent grain for the same slip system (see Fig. 1.7d). Furthermore, slip traces can be continuous on parts of a GB and be discontinuous on other parts of the same GB (Bridier et al., 2005; Abuzaid et al., 2012). Summarizing, one can state that the effective transmission and activation behavior of dislocations at GBs is a result of dislocation reactions (Lee et al., 1989a; Medlin et al., 1997), the type of dislocations involved (Zghal et al., 2001; Zghal and Couret, 2001), and the microstructural characteristics such as the GB type (Lim and Raj, 1985a; Gemperlova et al., 2004; Gemperle et al., 2005; Pond et al., 2006) or phase composition in multi-phase materials (Takasugi et al., 1978; Forwood and Clarebrough, 1981; De Hosson et al., 2006).

1.4 An overview on slip transmission criteria

1.4.1 Motivation

In the literature, criteria have been proposed to predict the dislocation transmission and activation behavior of slip systems in the grains adjacent to GBs. Continuum theories that would be able to model such phenomena need to account for these criteria that, for example, consider the orientation of slip systems and of the GB, respectively. Therefore, at first, an overview on these criteria is given with the focus on works considering experimental data. Subsequently, in Section 6.2, this overview is supplemented by a discussion of the literature on continuum crystal plasticity models considering slip transmission criteria. Finally, all geometric criteria are compared in a single-slip setting in Section 6.4. The following overview is taken from Bayerschen et al. (2016a).

1.4.2 Criteria that account for slip system orientations

Livingston and Chalmers (1957) were among the first to use geometric slip transmission criteria in experiments to predict the activated slip systems in a grain adjacent to a grain featuring a dislocation pile-up. Their criterion accounts for the orientations of the slip directions $\boldsymbol{d}_\alpha^\mathrm{A}, \boldsymbol{d}_\beta^\mathrm{B}$, and the orientations of the slip plane normals $\boldsymbol{n}_\alpha^\mathrm{A}, \boldsymbol{n}_\beta^\mathrm{B}$ (Fig. 1.8).

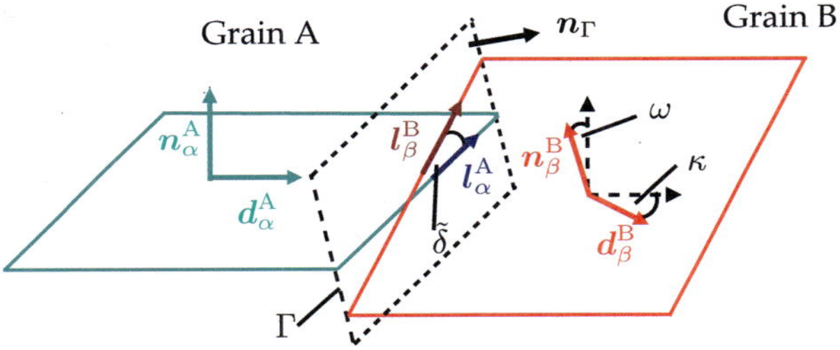

Figure 1.8: Nomenclature for slip systems α, β in adjacent grains A, B, separated by a grain boundary Γ. Figure reprinted from Bayerschen et al. (2016a) with permission from Springer.

Their used geometric transmission factor matrix for this purpose reads

$$\hat{N}_{\alpha\beta} = (\boldsymbol{n}_\alpha^\mathrm{A} \cdot \boldsymbol{n}_\beta^\mathrm{B})(\boldsymbol{d}_\alpha^\mathrm{A} \cdot \boldsymbol{d}_\beta^\mathrm{B}) + (\boldsymbol{n}_\alpha^\mathrm{A} \cdot \boldsymbol{d}_\beta^\mathrm{B})(\boldsymbol{n}_\beta^\mathrm{B} \cdot \boldsymbol{d}_\alpha^\mathrm{A}), \tag{1.4}$$

and has $N \times N$ components. With this criterion, the activation stress of the outgoing slip system is calculated (Livingston and Chalmers, 1957). It is based on the approximation that the stress state in the adjacent grain, resulting from a dislocation pile-up on the incoming slip system at the GB, is of pure shear stress type. Thus, the shear stresses on the incoming and on the outgoing slip systems are interconnected by the individual transmission factors. This purely geometric criterion is also used in Davis et al. (1966), following the interpretation of an activation of dislocation sources due to pile-ups.

A slightly modified version of the geometric transmission factor matrix (1.4) is employed in Luster and Morris (1995). The second term of (1.4) is dropped, and the transmission factor, therefore, reads

$$\hat{N}^{\text{mod}}_{\alpha\beta} = (\boldsymbol{n}^{\text{A}}_{\alpha} \cdot \boldsymbol{n}^{\text{B}}_{\beta})(\boldsymbol{d}^{\text{A}}_{\alpha} \cdot \boldsymbol{d}^{\text{B}}_{\beta}). \tag{1.5}$$

In the transmission evaluation of Guo et al. (2014), this factor is combined with the Schmid factors (Schmid and Boas, 1935) and a stress-intensity factor resulting from pile-ups (based on Eshelby et al., 1951). It is found, there, that a lower stress-intensity factor (leading to a lower resolved shear stress) on the emission slip system correlates to larger RBVs. For controlling the slip system activation, good alignment of slip systems has proved to be more important than a high Schmid factor. The importance of misalignment of slip systems for slip transmission processes has also been demonstrated with the preceding geometric factor in micro-hardness measurements of GBs (Wo and Ngan, 2004).

1.4.3 Criteria that account for slip system orientations and grain boundary orientation

In Shen et al. (1986), using (1.4) for the activation prediction is compared to using a different criterion incorporating the GB orientation via

$$\hat{M}_{\alpha\beta} = (\boldsymbol{l}^{\text{A}}_{\alpha} \cdot \boldsymbol{l}^{\text{B}}_{\beta})(\boldsymbol{d}^{\text{A}}_{\alpha} \cdot \boldsymbol{d}^{\text{B}}_{\beta}), \tag{1.6}$$

where $\boldsymbol{l}^{\text{A}}_{\alpha}, \boldsymbol{l}^{\text{B}}_{\beta}$ are normalized vectors of the lines of intersection, see Fig. 1.8. They can be obtained from, e.g., $\boldsymbol{l}^{\text{A}}_{\alpha} = (\boldsymbol{n}^{\text{A}}_{\alpha} \times \boldsymbol{n}_{\Gamma})/|(\boldsymbol{n}^{\text{A}}_{\alpha} \times \boldsymbol{n}_{\Gamma})|$. Here, $\boldsymbol{n}_{\Gamma}$ denotes the GB normal. In combination with a stress criterion, based on maximizing the Peach-Koehler force on the emitted dislocation, (1.6) is shown to predict all slip system activations, whereas the geometric criterion (1.4) did not. The geometric criterion (1.6) determines the slip plane, and the stress criterion determines the slip

direction of the emitted dislocation. These criteria are also used to predict the activation of slip systems in Shen et al. (1988).
The criteria for slip transmission are further extended by Lee et al. (1989b) to account for the RBV, where

$$\hat{M}^{\mathrm{mod}}_{\alpha\beta} = \boldsymbol{l}^{\mathrm{A}}_{\alpha} \cdot \boldsymbol{l}^{\mathrm{B}}_{\beta} \tag{1.7}$$

is used instead of (1.6). At first, the slip plane for a possible transmission is found by the slip plane normal that maximizes the scalar product between the lines of intersection. This corresponds to a minimization of the angle $\tilde{\delta}$, see "1." in Fig. 1.9. Then, the slip direction on this slip plane is determined by finding the maximum resolved shear stress (RSS) on the outgoing slip directions (see "2." in Fig. 1.9). In case of multiple slip directions with similar RSS, the direction is chosen which minimizes the RBV $\boldsymbol{b}_{\mathrm{r}}$ left behind in the GB after a transmission event. Thus, the angle κ between the slip directions is minimized (see "3." in Fig. 1.9). This approach removed remaining inconsistencies highlighted in the approach of Shen et al. (1988), and is used by Abuzaid et al. (2012). In Lee et al. (1990), it is proposed that the criteria of maximum RSS and minimum RBV need to be combined as they are competitive in nature (see also Lim and Raj, 1985b). It is found, however, that minimizing the RBV is of dominant influence for the slip transmission events. In Clark et al. (1992), it is outlined, with reference to Bamford et al. (1988), that this combined criterion is not applicable to multiple active slip systems. For the case of intermetallic phase boundaries, the combined criteria by Lee et al. (1990) have been shown to be applicable upon a slight refinement for metals with multiple types of slip systems (Misra and Gibala, 1999). The purely geometric criterion (1.6) is applied in Soer and De Hosson (2005), while in Tiba et al. (2015), it is applied and combined with investigations considering the RBV criterion and the incompatibility stresses. The importance of the RBV for slip transmission prediction is emphasized in Patriarca et al. (2013), as well.

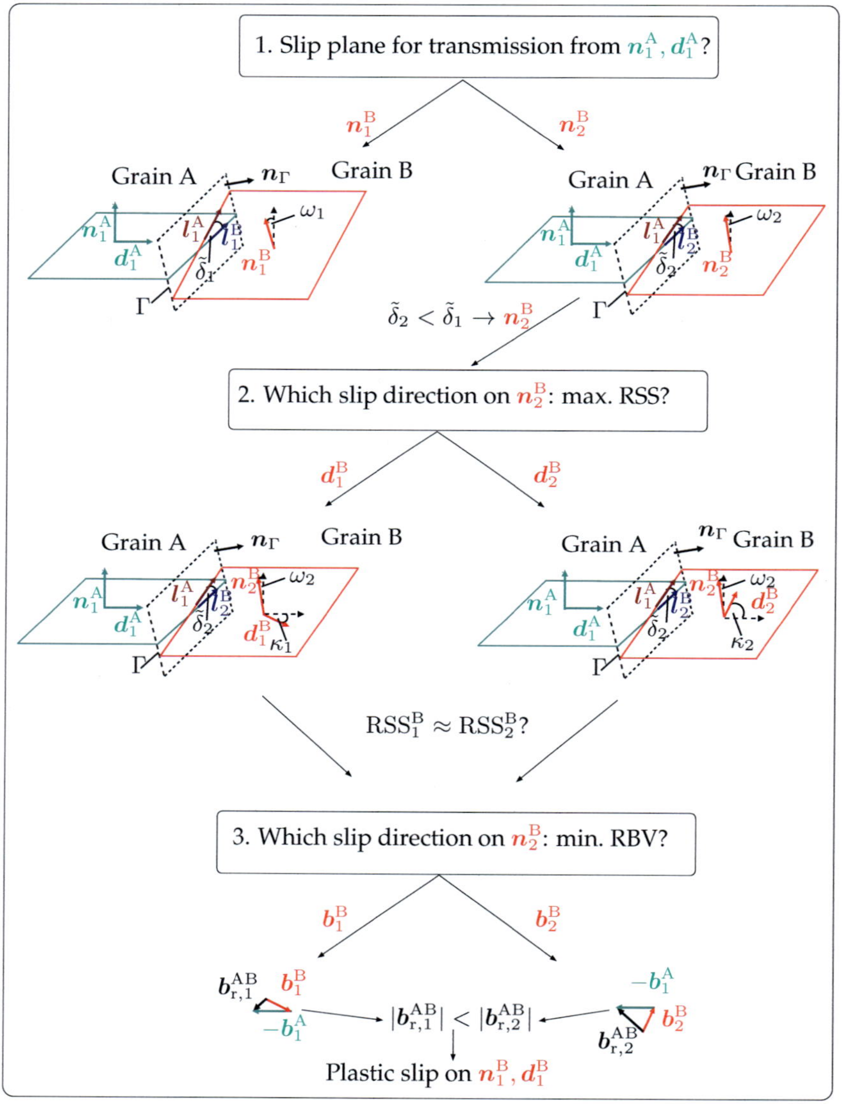

Figure 1.9: Combined criteria by Lee et al. (1989b) for slip transmission from grain A to B, separated by a grain boundary Γ. Figure reprinted from Bayerschen et al. (2016a) with permission from Springer.

1.4.4 Criteria that consider threshold values for the slip system and grain boundary angles

The geometrical criteria outlined in the previous sections were evaluated for each slip system, individually, to determine the most likely system for slip transfer across GBs. However, it has also been proposed in the literature to calculate an overall measure of slip transfer by, e.g., summing over the individual components of all possible slip system combinations. Such an approach is used in Werner and Prantl (1990). The mismatch between slip systems in adjacent grains is taken into account via a sum of the form

$$\tilde{\lambda} = \sum_{\alpha=1}^{N}\sum_{\beta=1}^{N} \cos\left(\frac{90^\circ}{\omega_c}\arccos\left(\boldsymbol{n}_\alpha^{\mathrm{A}} \cdot \boldsymbol{n}_\beta^{\mathrm{B}}\right)\right) \cos\left(\frac{90^\circ}{\kappa_c}\arccos\left(\boldsymbol{d}_\alpha^{\mathrm{A}} \cdot \boldsymbol{d}_\beta^{\mathrm{B}}\right)\right). \tag{1.8}$$

Here, ω_c and κ_c are critical angles above which no slip transfer is expected to occur on the associated slip systems. Thus, slip system combinations featuring a higher shared angle are not considered and removed from the sum. As it is discussed in Werner and Prantl (1990), the mismatch between slip plane normals is taken into account, rather than the mismatch between lines of intersection on Γ. This is due to the fact that the GB orientation was difficult to measure. The critical angles, above which slip transmission is not expected to occur, are taken to be $\kappa_c = 45^\circ$ and $\omega_c = 15^\circ$ for α/α- and α/β-phase boundaries in brass. For β/β-phase boundaries $\omega_c = 30^\circ$ is used as critical angle for the slip plane normals. The limit angle $\omega_c = 15^\circ$ was motivated by the work of Davis et al. (1966) using (1.4), where the critical angle $\tilde{\delta}_c$ between the lines of intersection is estimated to be in the range of $10^\circ - 20^\circ$. Furthermore, it is argued that the angle of lines of intersection for a pair of slip systems on adjacent sides of a GB cannot exceed the angle between adjacent slip plane normals, i.e., $\tilde{\delta} \leq \omega \rightarrow \boldsymbol{l}_\alpha^{\mathrm{A}} \cdot \boldsymbol{l}_\beta^{\mathrm{B}} \leq \boldsymbol{n}_\alpha^{\mathrm{A}} \cdot \boldsymbol{n}_\beta^{\mathrm{B}}$. Thus, ω is used in place

of $\tilde{\delta}$, see Fig. 1.8. This approach predicted the behavior of phase- / GBs with regard to their slip permeability.
In Kumar (2010), however, (1.8) is used in combination with the Schmid factors to investigate both criteria regarding the tensile strength of the considered material. It is found that the trend of the tensile strength is opposite to that of the calculated transmission number $\tilde{\lambda}$, i.e., a high value of $\tilde{\lambda}$ does not lead to an increased yield strength.
In Beyerlein et al. (2012), the angle $\tilde{\delta}$ between the lines of intersection is taken into account, rather than the angle ω between the slip plane normals. Instead of the summation in (1.8), individual components

$$\hat{\chi}_{\alpha\beta} = \cos\left(\frac{90^\circ}{\tilde{\delta}_\mathrm{c}}\arccos\left(\boldsymbol{l}_\alpha^\mathrm{A} \cdot \boldsymbol{l}_\beta^\mathrm{B}\right)\right)\cos\left(\frac{90^\circ}{\kappa_\mathrm{c}}\arccos\left(\boldsymbol{d}_\alpha^\mathrm{A} \cdot \boldsymbol{d}_\beta^\mathrm{B}\right)\right) \qquad (1.9)$$

are considered. The same critical angles as in the previous works, however, are utilized. This geometrical criterion is combined with the Schmid factors and further considerations regarding the interface shear strength (Demkowicz and Thilly, 2011; Wang et al., 2011; 2012).

1.4.5 Criteria that consider weighted sums of geometric transmission factors

Besides (1.8), other summation approaches exist in the literature considering, additionally, weights for the slip system contributions. Such an approach is taken in Bieler et al. (2014) since no clear correspondence to the transmission events could be established using the geometric factor (1.5) without weighting. Several weighted sum approaches for a slip transmission factor are proposed in Bieler et al. (2014). These scalar measures are based on the above described geometric factors. They are obtained by the summation over all slip system transmission factors and by weighting each one with plastic slips γ_α^A or the Schmid factors $m_\alpha^{\mathrm{A}}, m_\beta^{\mathrm{B}}$. Two such criteria are proposed in Bieler et al. (2014) by

$$m'_m = \sum_{\alpha,\beta} \hat{N}_{\alpha\beta}^{\mathrm{mod}} m_\alpha^{\mathrm{A}} m_\beta^{\mathrm{B}} \Big/ \sum_{\alpha,\beta} m_\alpha^{\mathrm{A}} m_\beta^{\mathrm{B}}, \tag{1.10a}$$

$$m'_\gamma = \sum_{\alpha,\beta} \hat{N}_{\alpha\beta}^{\mathrm{mod}} \gamma_\alpha^{\mathrm{A}} \gamma_\beta^{\mathrm{B}} \Big/ \sum_{\alpha,\beta} \gamma_\alpha^{\mathrm{A}} \gamma_\beta^{\mathrm{B}}. \tag{1.10b}$$

Measure (1.10a) connects the geometric mismatch with the RSS due to the employed weighting using Schmid factors. Two other measures are also given in Bieler et al. (2014) by

$$s_\gamma = \sum_{\alpha,\beta} \hat{M}_{\alpha\beta}^{\mathrm{mod}} \hat{N}_{\alpha\beta}^{\mathrm{mod}} \gamma_\alpha^{\mathrm{A}} \gamma_\beta^{\mathrm{B}} \Big/ \sum_{\alpha,\beta} \gamma_\alpha^{\mathrm{A}} \gamma_\beta^{\mathrm{B}}, \tag{1.11a}$$

$$LRB_\gamma = \sum_{\alpha,\beta} \hat{M}_{\alpha\beta} \gamma_\alpha^{\mathrm{A}} \gamma_\beta^{\mathrm{B}} \Big/ \sum_{\alpha,\beta} \gamma_\alpha^{\mathrm{A}} \gamma_\beta^{\mathrm{B}}. \tag{1.11b}$$

For the sample investigated in Bieler et al. (2014), all four measures give similar distributions along the GBs. This raises the question if the weighting by resolved shear stresses or by plastic slips is applicable. Furthermore, in (1.11a) the slip system normals seem to be double-accounted for due to the combination of both the geometric factors (1.5) and (1.7). The slip transmission criteria used in experiments are summarized in Table A.1.

1.5 Size effects of crystalline materials

It is known from experiments that the overall mechanical response of crystalline materials can be altered by changing the size of the grains, for example, by using heat treatments. The first observations and explanations of this phenomenon go back to works by Hall (1951b); Petch (1953). Therefore, this effect is referred to as the Hall-Petch effect. More recent investigations include the works by Fleck et al. (1994); Yang et al.

(2012); Chen et al. (2015). One basic observation in all theses works is that a refinement of the grain size leads to an increase in the material yield strength, see Fig. 1.10a. Thus, the overall stress that is necessary to plastically deform the material increases.

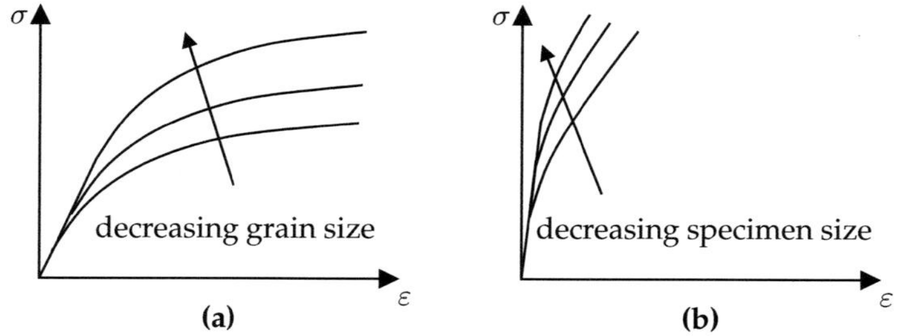

Figure 1.10: (a) Increase in overall yield strength of the stress response σ for specimens of identical size but decreasing grain size and applied strain ε. (b) Increase in overall yield strength of the stress response σ for specimens with decreasing specimen size but for identical grain size and applied strain ε.

The grain-size effect should be distinguished from the specimen size effect, i.e., a decrease in the size of the specimen that leads to an increase in the yield strength, although the grain size is kept constant (see Fig. 1.10b). Consequently, the yield strength depends on the ratio of grain size to specimen size. This has systematically been shown for gold microwires, e.g., by Chen and Ngan (2011). Both of the above-mentioned size effects are well established phenomena in the micron regime. In the nanometer regime, however, it has been observed that the effect of a grain size refinement is inverted. This leads to a decrease in the yield strength of the material for decreasing grain size, see, e.g., Zhao et al. (2003).

The occurrence of the size effects in the micron regime can be explained as follows. Microstructural lengths such as the distance to the next GB restrict the mean free path of the dislocations. These lengths are commonly referred to as characteristic lengths or internal lengths. When, for example, the grain size is refined for a constant specimen size and a constant dislocation density, the dislocations have less space to move through the lattice. Consequently, the tendency of dislocations to build pile-ups at GBs is increased. Therefore, other dislocations are progressively impeded by these dislocation pile-ups, enhancing the impediment of the dislocations to move. This can be interpreted as an increase in the local yield strength of the material. In consequence of this local increase, the overall yield strength of the material is increased, too.

1.6 The internal length scale of crystalline materials with dislocations

Associated to the dislocation pile-ups and other microstructural material characteristics, an internal length scale for the movement of dislocation exists. This internal length can be described as the effective distance between dislocations and the obstacles impeding their motion. For example, in a single-crystalline material region without the influence by GBs and defects other than dislocations, this distance is the mean spacing between the dislocations. If, however, the material is composed such that obstacles such as GBs, precipitates, or dislocation sources affect the motion of the dislocations, then the characteristic length is determined by the smallest distance of dislocations to the next obstacle. Therefore, this length scale arising also due to the presence of defects is, in general, not necessarily constant. In fact, it depends on influences such as the specimen size and the grain size, the dislocation density and the density distribution, the density of other defects and the dislocation source lengths (see Zhang et al., 2014).

1.7 Crystallographic texture influence on the mechanical response of crystalline materials

Due to the crystal lattice structure of materials, featuring preferred planes for plastic slip to take place, the elastic as well as the plastic mechanical response of many materials is orientation-dependent. Depending on the orientation of the loading with respect to the slip systems, slip systems can be oriented more or less favorable for an activation. For example, if the slip systems depicted in Fig. 1.5b are considered, loading along one of the depicted slip directions $\boldsymbol{d}$ leads to a high likeliness of an activation of plastic slip into this direction. The other slip directions are less likely to be activated. For the multi-slip case, the activation of slip systems depends on the orientation of all slip systems. If the crystal orientation is adjusted such that none of the slip directions are oriented similarly as the loading direction, plastic slip is impeded and the mechanical response strengthens. The resulting strengthening effect achieved by this adjustment is called texture strengthening. This effect is most pronounced for single-crystalline materials, see Fig. 1.11a for an exemplary single-crystalline material region, exhibiting only one crystal lattice orientation.

(a) (b) (c)

Figure 1.11: Schematic: (a) Single-crystalline material. Crystal lattice orientation illustrated in cut-free region. (b) Oligocrystalline material. (c) Polycrystalline material. Dashed lines in (b) and (c) illustrate grain boundaries.

Commonly, texture strengthening is utilized, for example, in applications such as high-pressure tubing or turbine blades. Metals usually possess a polycrystalline microstructure, see Fig. 1.11c for an exemplary polycrystalline material region (the different crystal orientations are indicated by colors). In this case, the orientation of the crystal lattice varies among the grains present in the material. Depending on the production process, these grains may, for example, be oriented arbitrarily without a preferred orientation, in which case the material responds effectively isotropic. Then, the mechanical response is not altered if the loading direction is changed. Thus, the mechanical material behavior is orientation-independent. However, in applications such as sheet metals, the production steps (e.g., cold forming) usually lead to a textured microstructure. In this case, the mechanical material response can depend largely on the applied loading direction, i.e., the material behavior is anisotropic.

The grain aggregates considered in the present work are mostly of oligocrystalline microstructure. In Fig. 1.11b, such a microstructure is depicted. It shows characteristics in between single-crystalline and polycrystalline materials. Several grains are present with different crystal orientations. However, the grains are often larger, compared to polycrystalline materials. Due to the presence of only a few grains with different crystal orientations, this type of microstructure may be substantially anisotropic. Compared to the polycrystalline case with random crystal orientations, the orientation spread of the grains may also be less pronounced. Thus, proper consideration of the influence of the crystal orientations has to be assured in modeling the material behavior of such grain aggregates.

1.8 Continuum representation of dislocation-induced plasticity

As shown in the previous sections, the motion of dislocations is a discrete process. For a sufficiently large amount of dislocations in the material, continuum theories can be used to describe the material behavior.

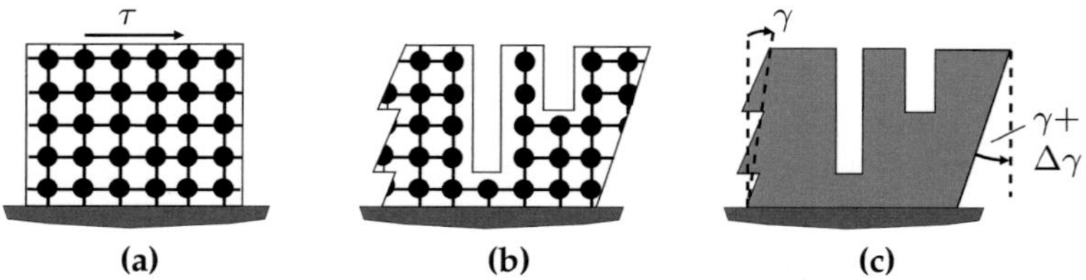

Figure 1.12: (a) Undeformed single-crystalline material region with shear load τ. (b) Deformed material region due to dislocation movement, elastic deformation not depicted. (c) Plastic slip γ in continuum representation of the deformation of the material region.

Therefore, a continuum representation of the local, discrete plastic slip is necessary. The dislocations are then not resolved individually, anymore. An initially undeformed, single-crystalline material region is considered, constrained by a fixed bearing at the bottom. On the top, it is loaded by a constant shear stress τ (see Fig. 1.12a). Dislocations may enter the material on the surface and move through the lattice, driven by the shear stress τ. This results in the creation of surface steps, see Fig. 1.12b for the elastically undeformed structure. On the continuum scale, however, such surface steps are not resolved. The plastic shearγ is defined as an average, smoothed quantity, see Fig. 1.12c. The movement of several dislocations of the same sign into the material leads to a non-constant field distribution of the continuum plastic shear γ. Therefore, gradients $\nabla\gamma$ occur in the plastic shear distribution. These can be considered in the free energy formulation of the material model, as it is shown, subsequently.

The Burgers vector can be defined on the continuum scale, too. Therefore, the following definition of the plastic distortion is considered

$$\boldsymbol{H}^{\mathrm{p}} = \sum_{\alpha=1}^{N} \gamma_\alpha \boldsymbol{d}_\alpha \otimes \boldsymbol{n}_\alpha, \tag{1.12}$$

where $\alpha = 1, \ldots, N$ denote the different slip systems and N is the number of slip systems which depends on the crystal lattice structure. For the depicted single-slip case, $\alpha = 1$, and, therefore, Eq. (1.12) reduces to

$$\boldsymbol{H}^{\mathrm{p}} = \gamma \boldsymbol{d} \otimes \boldsymbol{n}. \tag{1.13}$$

The total Burgers vector related to the plastic distortion reads

$$\boldsymbol{b}^{\mathrm{tot}} = \oint_{\mathcal{C}} \boldsymbol{H}^{\mathrm{p}} \, \mathrm{d}\boldsymbol{x}. \tag{1.14}$$

By application of Stoke's theorem, this can be reformulated as

$$\boldsymbol{b}^{\mathrm{tot}} = \int_{A} \mathrm{curl}\,(\boldsymbol{H}^{\mathrm{p}})^{\mathsf{T}} \, \mathrm{d}\boldsymbol{a} \tag{1.15}$$

with the rotation $\mathrm{curl}\,(\boldsymbol{B}) = \epsilon_{ijk} \partial B_{li} / \partial \boldsymbol{x}_k \boldsymbol{e}_l \otimes \boldsymbol{e}_j$, and A, the area bounded by $\mathcal{C}$, as well as $\mathrm{d}\boldsymbol{a} = \boldsymbol{n}_A \, \mathrm{d}a$. The quantity

$$\boldsymbol{\alpha} = \mathrm{curl}\,(\boldsymbol{H}^{\mathrm{p}})^{\mathsf{T}} \tag{1.16}$$

is usually referred to as a Burgers vector density per unit area, or simply as a dislocation density tensor (Nye, 1953).

For the single slip case from Fig. 1.12, $\boldsymbol{\alpha}$ is given by $\boldsymbol{\alpha} = -\boldsymbol{d} \cdot \nabla \gamma \boldsymbol{d} \otimes \tilde{\boldsymbol{l}}$. The scalar quantity $-\boldsymbol{d} \cdot \nabla \gamma$ offers the interpretation of a dislocation density. In the particular case depicted above, these dislocations are purely of edge-type since $\boldsymbol{d} \perp \tilde{\boldsymbol{l}}$. This density is often denoted by $\rho_{\vdash}$.

The components of $\boldsymbol{\alpha}$ with $\boldsymbol{d} \parallel \tilde{\boldsymbol{l}}$ are screw dislocation densities $\rho_{\odot}$.
In many continuum models, the dislocation densities of edge and screw dislocations are used to describe the plastic deformation, e.g., Gurtin et al. (2007); Van Beers et al. (2013). For the multi-slip case, the resulting numerical effort is comparably high.

1.9 Crystal plasticity

1.9.1 Basic assumptions

In the following, the key ingredients of crystal plasticity to describe geometrically linear dislocation-induced plasticity are outlined. For simplicity, the single slip case is considered. In the subsequent sections, however, the theory is refined for the multi-slip case.
The displacement gradient, $\boldsymbol{H} = \mathrm{grad}\,(\boldsymbol{u})$, is assumed to be additively decomposable into an elastic part $\boldsymbol{H}^{\mathrm{e}}$ and a plastic part $\boldsymbol{H}^{\mathrm{p}}$ by

$$\boldsymbol{H} = \boldsymbol{H}^{\mathrm{e}} + \boldsymbol{H}^{\mathrm{p}}. \tag{1.17}$$

From the definition of the infinitesimal strain tensor

$$\boldsymbol{\varepsilon} = \mathrm{sym}(\boldsymbol{H}) = \frac{1}{2}(\boldsymbol{H} + \boldsymbol{H}^{\mathsf{T}}) \tag{1.18}$$

and Eq. (1.17) it follows that

$$\boldsymbol{\varepsilon} = \boldsymbol{\varepsilon}^{\mathrm{e}} + \boldsymbol{\varepsilon}^{\mathrm{p}}. \tag{1.19}$$

For the single slip case, by Eq. (1.12) and Eq. (1.19), the plastic strain reads $\boldsymbol{\varepsilon}^{\mathrm{p}} = \mathrm{sym}(\gamma \boldsymbol{d} \otimes \boldsymbol{n})$. The relation between the elastic part of the infinitesimal strain tensor $\boldsymbol{\varepsilon}^{\mathrm{e}}$ and the Cauchy stress $\boldsymbol{\sigma}$ is assumed to be given by Hooke's law

$$\boldsymbol{\sigma} = \mathbb{C}[\boldsymbol{\varepsilon}^{\mathrm{e}}], \tag{1.20}$$

where $\mathbb{C}$ denotes the fourth-order elastic stiffness tensor. Then, the elastic energy density (per volume) can be written as

$$W_{\mathrm{e}} = \frac{1}{2}\boldsymbol{\sigma} \cdot \boldsymbol{\varepsilon}^{\mathrm{e}}. \tag{1.21}$$

The bulk stress power density p_{vol} expended by the Cauchy stress reads

$$p_{\mathrm{vol}} = \boldsymbol{\sigma} \cdot \dot{\boldsymbol{\varepsilon}} = \boldsymbol{\sigma} \cdot (\dot{\boldsymbol{\varepsilon}}^{\mathrm{e}} + \dot{\boldsymbol{\varepsilon}}^{\mathrm{p}}) = \dot{W}_{\mathrm{e}} + \tau\dot{\gamma} \tag{1.22}$$

where the resolved shear stress τ on the considered slip plane is given by $\tau = \boldsymbol{\sigma} \cdot \mathrm{sym}(\boldsymbol{d} \otimes \boldsymbol{n})$. Consequently,

$$p_{\mathrm{vol}} = \dot{W}_{\mathrm{e}} + \dot{W}_{\mathrm{p}} \tag{1.23}$$

with $\dot{W}_{\mathrm{p}} = \tau\dot{\gamma}$. Obviously, the stress power density p_{vol} is expressible as the sum of the rate of the elastically stored energy density $\dot{W}_{\mathrm{e}}$ and a plastic power density contribution $\dot{W}_{\mathrm{p}}$. This is exploited further, for the model derivations, in the following chapters.

1.9.2 Incorporation of viscoplasticity

The incorporation of the plastic slip rate $\dot{\gamma}$ in p_{vol} motivates the definition of a constitutive law for the evolution of $\dot{\gamma}$. It is known that metals behave rate-dependent, i.e., depending on the loading-rate of the applied strain (e.g., Johnston, 1962; Taylor, 1965; Patel and Bieniek, 1979). A common shear rate law considering this type of behavior, for example, is the overstress-type formulation (see, e.g., Little et al., 1981; Sung et al., 2010, for an overview on rate-dependent formulations)

$$\dot{\gamma} = \dot{\gamma}_0 \,\mathrm{sg}(\tau) \left\langle \frac{|\tau| - \tau^{\mathrm{C}}}{\tau^{\mathrm{D}}} \right\rangle^{p}. \tag{1.24}$$

The Macauley-brackets ensure positiveness of their argument. For example, considering a scalar quantity a, $\langle a \rangle = \max(a, 0)$. The rate-dependence is realized through the rate-sensitivity parameter p, the critical resolved shear stress for an activation of the slip system is τ^{C}, and the quantity τ^{D} is called drag stress. The reference shear rate is denoted by $\dot{\gamma}_0$. All of these four quantities are material parameters that are usually identified in experiments.
Although metals in general behave rate-dependent, for particular applications, such as slow deformation processes, rate-independent models can be used. For such cases, also the viscoplastic models can be applied in order to regularize the issues that may arise with rate-independent formulations and implementations (Han and Reddy, 2012).

1.9.3 Hardening due to dislocations

As mentioned in the beginning of this chapter, the plastic mechanical response of metals is not necessarily uniform but rather evolves with the proceeding deformation of the material. Therefore, the critical resolved shear stress τ^{C} is not a constant, but depends on the state of deformation and the concentration of defects. In general, it is also temperature-dependent, but these effects are neglected in the present work.
The Taylor-relation (Taylor, 1934) was one of the first formulations connecting the critical resolved shear stress to the plastic state of the material via the dislocation density ρ

$$\tau^{\mathrm{C}} = cbG\sqrt{\rho}, \tag{1.25}$$

where G is the shear modulus, b is the Burgers vector magnitude, and c is a constant of the order of unity, e.g., $c \approx 0.5 \ldots 1$. The relation (1.25) is also used in gradient-extended plasticity theories. For example, the theory by Nix and Gao (1998) considers an additive split of the dislocation density ρ into a GND-density contribution ρ_{GND} and an

SSD-density contribution ρ_{SSD}. This split results in a critical resolved shear stress of the following form

$$\tau^{\text{C}} = cbG\sqrt{\rho_{\text{GND}} + \rho_{\text{SSD}}}. \tag{1.26}$$

When the influence of GNDs can be neglected due to, for example, sufficiently homogeneous deformations, Eq. (1.26) reduces to

$$\tau^{\text{C}} = cbG\sqrt{\rho_{\text{SSD}}}. \tag{1.27}$$

On this account, it should also be noted that the influence of GNDs on the mechanical response can be modeled by other approaches, details of which are given in the next section.

Continuum models and their employed hardening relations are intended to be calibrated to experiments. Latent hardening models (e.g., Kocks and Brown, 1966; Lavrentev and Pokhil, 1975; Lavrentev, 1980; Franciosi and Zaoui, 1982) are commonly used (e.g., Asaro and Needleman, 1985; Anand and Kothari, 1996). They consider hardening occurring in other (than primary) slip systems due to slip in primary slip systems. The Mises-Hill framework of Gurtin and Reddy (2014) considers self-hardening (i.e., slip systems harden due to their own plastic slip) and latent hardening. A hardening rule is introduced on the individual slip systems in dependence of accumulated plastic slips. Other hardening approaches, based on scalar plastic strain measures, have been developed (see Bouvier et al., 2005, for a brief overview). Such approaches allow for rapid model parameter calibration with a manageable number of experiments.

In the following, $\gamma \geq 0$ is assumed, for simplicity. A phenomenological hardening approach considering the plastic slip can be formulated using the Voce-type law (Voce, 1948; 1955) with a hardening stress β

$$\tau^{\text{C}} = \tau_0^{\text{C}} + \beta. \tag{1.28}$$

A typical value for the initial critical resolved shear stress is $\tau_0^{\mathrm{C}} \approx 1\,\mathrm{MPa}$ for gold (Sachs and Weerts, 1930). This is commonly referred to as friction stress of the slip system, i.e., the stress necessary to activate plastic slip in the absence of hardening induced by the presence of other dislocations. The hardening stress in the Voce-formulation reads

$$\beta = (\tau_\infty^{\mathrm{C}} - \tau_0^{\mathrm{C}}) - (\tau_\infty^{\mathrm{C}} - \tau_0^{\mathrm{C}}) \exp\left(-\frac{\Theta\gamma}{(\tau_\infty^{\mathrm{C}} - \tau_0^{\mathrm{C}})}\right), \tag{1.29}$$

where τ_∞^{C} is the saturation stress, and Θ is the initial hardening modulus. The stress β models dislocation-induced hardening caused by the random trapping of dislocations in the bulk material. Thus, the contributions to hardening by the SSDs are accounted for. For $\gamma \to 0$, it follows that $\beta \to 0 \Rightarrow \tau^{\mathrm{C}} = \tau_0^{\mathrm{C}}$, and, for $\gamma \to \infty$, that $\beta \to \tau_\infty^{\mathrm{C}} - \tau_0^{\mathrm{C}} \Rightarrow \tau^{\mathrm{C}} = \tau_\infty^{\mathrm{C}}$ (see Fig. 1.13 for an exemplary plot). This hardening law is based on the assumption that the hardening behavior saturates for large strains. Therefore, the hardening stress cannot increase unlimited but reaches a saturation stress τ_∞^{C}. It is noted that, for the multi-slip case, γ in this formulation may be substituted, for example, by a measure of the overall plastic deformation such as, e.g., an accumulated plastic slip γ_{ac}.

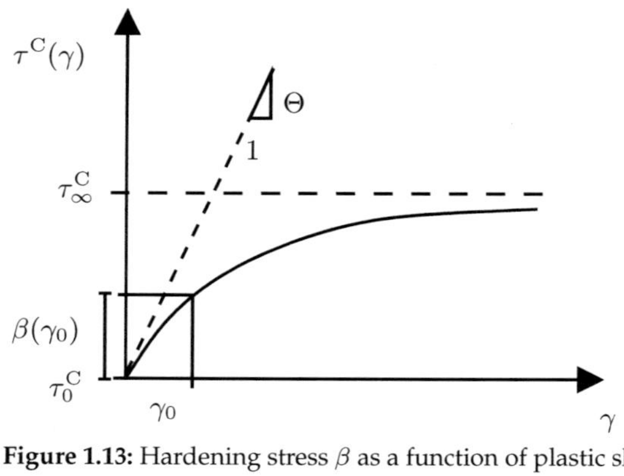

Figure 1.13: Hardening stress β as a function of plastic shear γ.

For monotonic loading processes, it can be shown that a direct formal connection exists between the Voce-law from Eq. (1.28), Eq. (1.29) and the dislocation-density-based Taylor formulation of hardening. Therefore, the following evolution law for the dislocation density is considered (Kocks, 1976)

$$\dot{\rho} = (k_1\sqrt{\rho} - k_2\rho)\dot{\gamma}, \tag{1.30}$$

where the first term models storage of dislocations and the second term accounts for annihilation of dislocations. The parameter k_1 is a constant, and $k_2 = k_2(\dot{\gamma})$ is a rate-dependent annihilation parameter. For the single slip case, the dislocation density ρ corresponds to the density of dislocations of this single slip system. Considering the case of constant shear rates $\dot{\gamma} = const.$, Eq. (1.30) can be reformulated by

$$\frac{\mathrm{d}\rho}{\mathrm{d}t} = (k_1\sqrt{\rho} - k_2\rho)\frac{\mathrm{d}\gamma}{\mathrm{d}t}, \tag{1.31}$$

or, equivalently, by

$$\mathrm{d}\rho = (k_1\sqrt{\rho} - k_2\rho)\,\mathrm{d}\gamma. \tag{1.32}$$

The derivative of the critical resolved shear stress, Eq. (1.25), with respect to the dislocation density reads

$$\frac{\mathrm{d}\tau^{\mathrm{C}}}{\mathrm{d}\rho} = \frac{cbG}{2\sqrt{\rho}}. \tag{1.33}$$

Additional substitution of Eq. (1.32) in Eq. (1.33) and regrouping of the terms gives

$$\frac{\mathrm{d}\tau^{\mathrm{C}}}{\mathrm{d}\gamma} = \frac{cbG}{2}(k_1 - k_2\sqrt{\rho}). \tag{1.34}$$

Substitution of the Taylor-relation Eq. (1.25) into Eq. (1.34) yields

$$\frac{\mathrm{d}\tau^{\mathrm{C}}}{\mathrm{d}\gamma} = \frac{cbG}{2}\left(k_1 - k_2\frac{\tau^{\mathrm{C}}}{cbG}\right). \tag{1.35}$$

Using the abbreviations $\tilde{\Theta} = cbGk_1/2$ and $\tau^{\mathrm{C}}_{\infty} = k_1 cbG/k_2$ in Eq. (1.35) leads to the Voce-form of Bouaziz (2012)

$$\frac{\mathrm{d}\tau^{\mathrm{C}}}{\mathrm{d}\gamma} = \tilde{\Theta}\left(1 - \frac{\tau^{\mathrm{C}}}{\tau^{\mathrm{C}}_{\infty}}\right). \tag{1.36}$$

After separation of variables, the integration of Eq. (1.36) gives

$$\frac{1}{\tilde{\Theta}}\ln\left(1 - \frac{\tau^{\mathrm{C}}}{\tau^{\mathrm{C}}_{\infty}}\right)(-\tau^{\mathrm{C}}_{\infty}) = \gamma + C. \tag{1.37}$$

The integration constant C is obtained from the initial condition for $\tau^{\mathrm{C}}(\gamma = 0) = \tau^{\mathrm{C}}_{0}$, and, thus,

$$C = \frac{1}{\tilde{\Theta}}\ln\left(1 - \frac{\tau^{\mathrm{C}}_{0}}{\tau^{\mathrm{C}}_{\infty}}\right)(-\tau^{\mathrm{C}}_{\infty}). \tag{1.38}$$

Combining Eq. (1.37) with Eq. (1.38) and solving for τ^{C} gives

$$\tau^{\mathrm{C}} = \tau^{\mathrm{C}}_{\infty} - (\tau^{\mathrm{C}}_{\infty} - \tau^{\mathrm{C}}_{0})\exp\left(-\frac{\tilde{\Theta}\gamma}{\tau^{\mathrm{C}}_{\infty}}\right). \tag{1.39}$$

Redefining the initial hardening modulus by $\tilde{\Theta} = \Theta\tau^{\mathrm{C}}_{\infty}/(\tau^{\mathrm{C}}_{\infty} - \tau^{\mathrm{C}}_{0})$ in Eq. (1.39) recovers the Voce-hardening law from Eq. (1.28), Eq. (1.29)

$$\tau^{\mathrm{C}} = \tau^{\mathrm{C}}_{\infty} - (\tau^{\mathrm{C}}_{\infty} - \tau^{\mathrm{C}}_{0})\exp\left(-\frac{\Theta\gamma}{\tau^{\mathrm{C}}_{\infty} - \tau^{\mathrm{C}}_{0}}\right). \tag{1.40}$$

Thus, the phenomenological Voce-hardening relation can be motivated from the evolution-law of the dislocation density, Eq. (1.30), and the Taylor-relation for the critical resolved shear stress, Eq. (1.25), for monotonic loading processes which are considered throughout the present work.

1.9.4 Hardening due to geometrically necessary dislocations: a gradient plasticity approach

Motivated by the classical plasticity theories failing to model unconventional phenomena such as size effects, the influence of GNDs on the mechanical material behavior has been shown in experimental and theoretical investigations going back to the early works by Ashby (1970). Nowadays, gradient-extended plasticity theories are commonly used to model these effects.

Many gradient plasticity theories use Nye's dislocation density tensor, Eq. (1.16), in order to consider the influence of non-local deformations, i.e., the gradient-related contributions (e.g., Shizawa and Zbib, 1999; Acharya and Bassani, 2000; Menzel and Steinmann, 2000; Arsenlis et al., 2004; Gurtin et al., 2007). In the following, an exemplary single-slip simple shear problem is discussed to illustrate the effects of a typical gradient-related additional defect stress in the hardening model. The defect energy density can be postulated as a potential from which the gradient stress can be derived. This free energy contribution is often taken to be quadratic in the gradients. One form of it reads

$$W_{\mathrm{g}}(\nabla\gamma) = W_0 \left(\frac{|\nabla\gamma|}{g_0}\right)^2 . \tag{1.41}$$

Here, W_0 is the initial defect energy density. For the case of a defect energy being linear (in the dislocation density, instead of the plastic slip), this value is commonly taken to be $W_0 = \bar{\alpha}G$, (cf. Hirth and Lothe, 1982; Ohno and Okumura, 2007) with $\bar{\alpha} \approx 0.5 \ldots 1$ usually assumed to be a constant. The quantity g_0 denotes a normalization constant and can, thus, be related to the internal length scale that is introduced by the defect energy in the theory. In the case of a linear defect energy, this constant could be taken to be the inverse of the Burgers vector length, i.e., $g_0 = 1/b$ (see, e.g., Ohno and Okumura, 2007).

The total free energy density is assumed to be the sum of the classic, elastic energy Eq. (1.21) and the defect contribution Eq. (1.41) via

$$W(\boldsymbol{\varepsilon}, \gamma, \nabla\gamma) = W_{\mathrm{e}}(\boldsymbol{\varepsilon}, \boldsymbol{\varepsilon}^{\mathrm{p}}(\gamma)) + W_{\mathrm{g}}(\nabla\gamma). \tag{1.42}$$

By the principle of virtual power, the connection between a vectorial gradient-related defect stress $\boldsymbol{\xi}$ and the resolved shear stress may be derived in form of an additional balance equation. For brevity, this is not performed here, but in the subsequent sections for the three-dimensional case. It is noted that the resulting so-called microforce balance is given by

$$\tau^{\mathrm{d}} = \tau + \operatorname{div}(\boldsymbol{\xi}), \tag{1.43}$$

where τ^{d} is the dissipative shear stress of the slip system, under the assumption that the dissipation of the bulk is induced by

$$\mathcal{D} = \tau^{\mathrm{d}}\dot{\gamma}. \tag{1.44}$$

For the multi-slip case, it is assumed that the dissipation contributions of all slip system can be superimposed (Cermelli and Gurtin, 2002). The derivative of Eq. (1.41), w.r.t. the gradient of the plastic slip, reads

$$\boldsymbol{\xi} = \frac{\partial W_{\mathrm{g}}(\nabla\gamma)}{\partial \nabla\gamma} = 2\frac{W_0}{g_0^2}|\nabla\gamma|\frac{\nabla\gamma}{|\nabla\gamma|} = 2\frac{W_0}{g_0^2}\nabla\gamma. \tag{1.45}$$

For single-slip,

$$\boldsymbol{\xi}\cdot\boldsymbol{e}_x = \xi_x = 2\frac{W_0}{g_0}\left(\frac{\left|\frac{\mathrm{d}\gamma}{\mathrm{d}x}\right|}{g_0}\right). \tag{1.46}$$

Clearly, the gradient stress for this quadratic type of the defect energy is linear in the plastic slip gradient. A different (non-quadratic) approach is explored in Chapter 5. The divergence of the gradient stress reads

$$\operatorname{div}(\boldsymbol{\xi}) = \mathrm{d}\xi_x/\,\mathrm{d}x. \tag{1.47}$$

Consequently, the gradient stress contribution to the dissipative shear stress in Eq. (1.43) is given by

$$\frac{\mathrm{d}\xi_x}{\mathrm{d}x} = 2\frac{W_0}{g_0^2}\frac{\mathrm{d}\left|\frac{\mathrm{d}\gamma}{\mathrm{d}x}\right|}{\mathrm{d}x}, \tag{1.48}$$

where the right-hand side derivative can be expressed by

$$\frac{\mathrm{d}\left|\frac{\mathrm{d}\gamma}{\mathrm{d}x}\right|}{\mathrm{d}x} = \mathrm{sg}\left(\left|\frac{\mathrm{d}\gamma}{\mathrm{d}x}\right|\right)\frac{\mathrm{d}^2\gamma}{\mathrm{d}x\mathrm{d}x}. \tag{1.49}$$

A simple form of the flow rule may, for example, be postulated by

$$\dot{\gamma} = \dot{\gamma}_0\left\langle\frac{\tau^{\mathrm{d}} - \tau_0^{\mathrm{C}}}{\tau^{\mathrm{D}}}\right\rangle. \tag{1.50}$$

By substituting Eq. (1.43) and Eq. (1.47) into Eq. (1.50), this reads

$$\dot{\gamma} = \dot{\gamma}_0\left\langle\frac{\tau - \left(\tau_0^{\mathrm{C}} - \frac{\mathrm{d}\xi_x}{\mathrm{d}x}\right)}{\tau^{\mathrm{D}}}\right\rangle. \tag{1.51}$$

Thus, the critical resolved shear stress is given by

$$\tau^{\mathrm{C}} = \tau_0^{\mathrm{C}} - \frac{\mathrm{d}\xi_x}{\mathrm{d}x}. \tag{1.52}$$

It can be shown that, for the single-slip case, considering homogeneous boundary conditions for the plastic slip γ, an average shear $\bar{\gamma}$, and a single-crystalline body of width $2l$ with passivated surfaces, the following is obtained for the average critical shear stress

$$\bar{\tau}^{\mathrm{C}} = \bar{\tau}_0^{\mathrm{C}} + 6\frac{W_0}{l^3 g_0^2}\bar{\gamma}. \tag{1.53}$$

In Section 5.3 the general derivation is provided. The hardening term, stemming from the quadratic defect energy, is linear in the average plastic shear. In addition, the influence by the specimen size $2l$ (in this single-crystalline case also the grain size) and by the internal length scale $l_{\text{int}} = 1/g_0$ are exhibited. The smaller the grain size and / or the larger the internal length scale, the higher is the hardening stress in Eq. (1.53), see Fig. 1.14a. Ideal plastic behavior is recovered in the limit case of $l_{\text{int}} = 0$.

The additional stress contribution can be interpreted as a back-stress caused by the presence of GNDs, i.e., by the pile-ups of dislocations at GBs. This can be visualized by means of the spatial distribution of plastic slip. For the considered case of the quadratic defect energy, the resulting distribution of plastic slip is

$$\gamma(x) = \frac{\Delta\bar{\tau} g_0^2}{4W_0}\left(l^2 - |x|^2\right) \tag{1.54}$$

with an average shear stress difference of $\Delta\bar{\tau} = \bar{\tau}^{\text{C}} - \bar{\tau}_0^{\text{C}}$. The distribution of $\gamma(x)$ is schematically depicted in Fig. 1.14b, showing the influence by the specimen size $2l$ and by the internal length scale l_{int}.

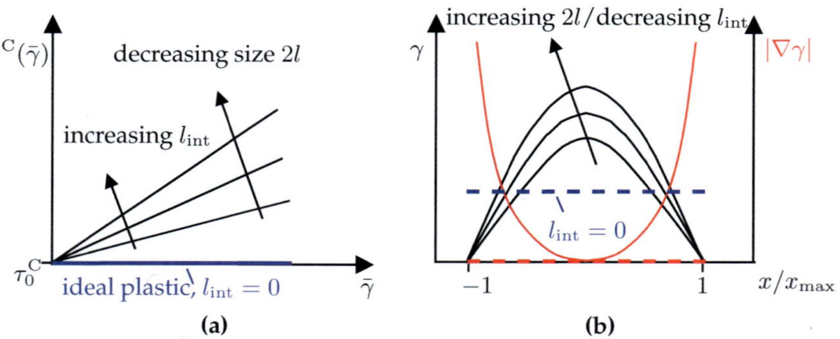

Figure 1.14: (a) Critical average shear stress $\bar{\tau}^{\text{C}}$ as a function of average plastic shear $\bar{\gamma}$ for a quadratic defect energy. (b) Corresponding spatial distributions of plastic slip.

Consistent with increased hardening, less pronounced plastic slip is obtained. For example, the smaller the internal length scale, the larger is the maximum plastic slip. The distribution of $\gamma(x)$ for the ideal-plastic material behavior is indicated by a dashed line (highlighted in blue). Higher hardening due to GNDs is associated to larger gradients of plastic slip. These are visualized (highlighted in red) for one of the slip distributions in Fig. 1.14b. For the ideal plastic response, no gradients are present (see dashed line in Fig. 1.14b, highlighted in red).
In real structures, the observed distributions of plastic slip are usually of more complex shape and not necessarily parabolic. Therefore, a non-quadratic generalization of the defect energy is proposed in Chapter 5.

1.9.5 Incorporation of grain boundary plasticity

In continuum theories, the collective dislocation behavior is modeled. Consequently, the mechanisms of dislocation interactions at GBs are also accounted for in an averaged manner. For the modeling of oligocrystalline aggregates, the incorporation of GB effects is of increased importance. Compared to polycrystalline aggregates, the mechanical response can be much more influenced by the characteristics of the GBs. On the continuum scale, the dislocations are modeled in an averaged fashion. Therefore, the discrete processes and reactions taking place at GBs cannot be distinguished anymore (e.g., Zhang et al., 2014). Instead, the GB dislocation processes are also modeled in an averaged sense. For example, on the continuum scale the resulting plastic slip field can be identical for two distinct discrete dislocation interactions. The direct transmission of a dislocation (see Fig. 1.6g) produces the same strain profile as the deposition of two dislocations at the GB (see Fig. 1.6d).
The incorporation of higher-order quantities such as the gradient stresses in gradient plasticity theories necessitate the formulation of additional boundary conditions. During the early developments of gradient plasticity (GP) theories, the boundaries and GBs were usually assigned

two types of idealized additional conditions. They were considered to either allow dislocation transfer unrestrictedly or not at all. The unrestricted-flow condition is termed microscopically free or simply microfree condition. In contrast, the non-passing condition is referred to as microhard condition (e.g., Gurtin, 2002). In continuum models, the microhard condition is imposed by, for example, setting the slip rates to zero near or on the GBs Γ

$$\dot{\gamma} = 0 \quad \forall x \in \Gamma. \tag{1.55}$$

The microfree case corresponds to a condition for the gradient stress

$$\boldsymbol{\xi} \cdot \boldsymbol{n} = 0 \quad \forall x \in \Gamma, \tag{1.56}$$

where $\boldsymbol{n}$ denotes the normal vector of Γ. This condition is formally similar to a classic macroscopic boundary condition for the Cauchy stress on the surface $\partial\mathcal{B}$

$$\boldsymbol{\sigma}\boldsymbol{n} = \boldsymbol{0} \quad \forall x \in \partial\mathcal{B}, \tag{1.57}$$

where $\boldsymbol{n}$ denotes the normal vector of $\partial\mathcal{B}$.

However, the two idealized conditions have been relaxed in recent years to allow for behavior in-between. Exemplary works exploring this are Van Beers et al. (2013); Wulfinghoff et al. (2013); Gottschalk et al. (2016). In these theories, an additional yield condition for the GBs is introduced. In conjunction with an assumed contribution of the GB plasticity to the free energy, a finite resistance of the GBs against plastic flow can be modeled. Details of such an approach are given in Chapter 3. In the present section, however, only an introductory overview on the behavior of these different types of boundary conditions is given. For example, two grains are considered with a single slip system each of nearly identical orientation. Initially, no dislocations are present in both grains. For simplicity, three dislocations in motion are considered, entering from the left in Fig. 1.15a (top), and highlighted in grey.

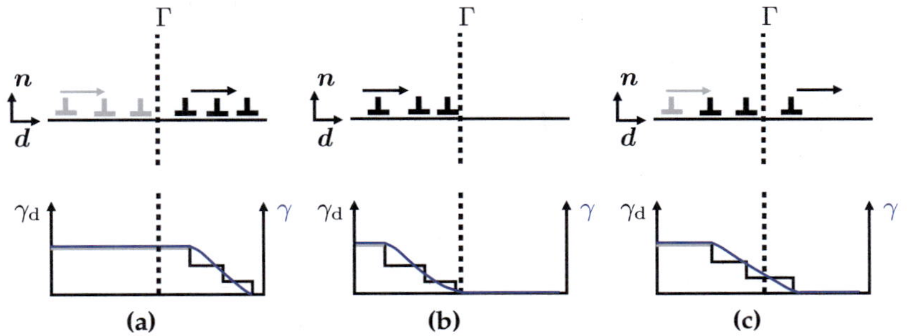

Figure 1.15: Top: Plastic flow of dislocations. Bottom: corresponding discrete and continuum plastic slip profiles. (a) Unrestricted by the grain boundary, (b) impenetrable grain boundary prohibiting dislocation movement, (c) intermediate grain boundary behavior, some dislocations can pass.

In the case of unrestricted plastic flow, all dislocations may pass the GB, as shown in Fig. 1.15a (top). The discrete slip distribution γ_d is indicated in black color, while the continuum slip distribution γ is highlighted in blue color in Fig. 1.15a (bottom). At the GB, the same value of γ can be observed as in the bulk material that the dislocations have already passed through. However, for the non-passing condition, Fig. 1.15b, the dislocations pile up at the GB and the plastic slip remains zero, there. For the case that the GB resistance against plastic flow is modeled as a finite value, some dislocations can pass and, thus, the pile-up at the GB is decreased (see Fig. 1.15c). The value of the plastic slip, there, is in between the values for the microfree and the microhard case.

A slightly different interpretation of the boundary/interface conditions can be given, when considering that dislocations are active on both sides of the GB. For the microfree condition, a homogeneous distribution of plastic slip is obtained (see, for example, Fig. 1.16a). In the case that the GB is microhard, the plastic slip remains zero, there, but plastic slip can be observed close to the GB in both grains, Fig. 1.16b. If the gradients of plastic slip are large enough, the GB yield condition is fulfilled, and the pile-ups are decreased. The plastic slip on the GB, then, takes non-zero values, see, e.g., Fig. 1.16c. The region over which the gradient influence is present in the continuum representation (schematically indicated by "g" in Fig. 1.16c) is determined by the internal length scale parameter. For this parameter, different interpretations exist to date (see Zhang and Aifantis, 2015, for an overview). For example, the value of the internal length scale of the gradient plasticity model by Aifantis and Willis (2004; 2005) was determined in nanoindentation studies close to a GB by Aifantis et al. (2006). There, the interpretation was given that this length was the distance over which 90% of the dislocations piled up.

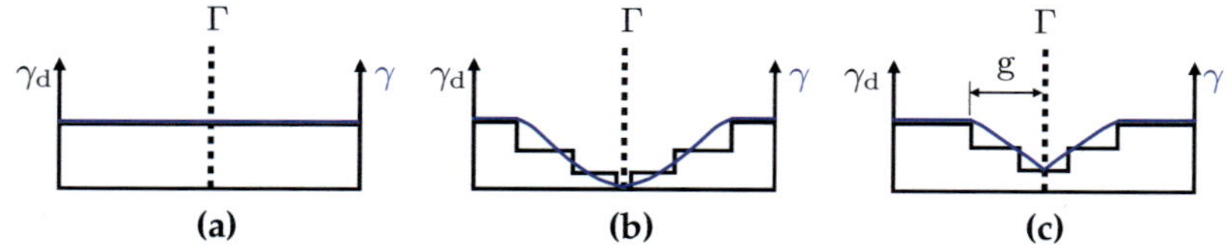

Figure 1.16: Discrete and continuum plastic slip profiles: (a) unrestricted by grain boundary, (b) impenetrable grain boundary, (c) some dislocations can pass the grain boundary. Region of gradient influence in the continuum plastic slip profile indicated by "g".

Chapter 2

Crystal plasticity with an accumulated plastic slip

2.1 Basic assumptions

Before grain boundaries and their influence, leading to inhomogeneous plastic deformations, are considered in the following chapters, a basic crystal-plasticity model is discussed, at first. A small-strain framework is used in which the displacement of a material point $\boldsymbol{x}$ of a body $\mathcal{B}$ is denoted by the vector $\boldsymbol{u}(\boldsymbol{x})$. The infinitesimal strain tensor is given by

$$\boldsymbol{\varepsilon} = \mathrm{sym}(\mathrm{grad}\,(\boldsymbol{u})), \tag{2.1}$$

with $\mathrm{sym}(\boldsymbol{A}) = \boldsymbol{A} + \boldsymbol{A}^{\mathsf{T}}$. An additive decomposition of the infinitesimal strain tensor is assumed by

$$\boldsymbol{\varepsilon}^{\mathrm{e}} = \boldsymbol{\varepsilon} - \boldsymbol{\varepsilon}^{\mathrm{p}}, \tag{2.2}$$

and the plastic part of the strain tensor reads

$$\boldsymbol{\varepsilon}^{\mathrm{p}} = \sum_{\alpha} \gamma_{\alpha} \mathrm{sym}(\boldsymbol{d}_{\alpha} \otimes \boldsymbol{n}_{\alpha}) \tag{2.3}$$

with the plastic slips γ_{α} of $\alpha = 1, \ldots, N$ slip systems. The slip directions are denoted by $\boldsymbol{d}_{\alpha}$ and the slip plane normals by $\boldsymbol{n}_{\alpha}$.

An accumulated plastic slip can be introduced, e.g., by considering the sum over all plastic slips on the individual.slip systems via

$$\gamma_{\mathrm{ac}}(\hat{\gamma}) = \sum_{\alpha} \int |\dot{\gamma}_\alpha| \, \mathrm{d}t = \sum_{\alpha} |\gamma_\alpha|, \tag{2.4}$$

For practical reasons, however, plastic slip parameters λ_α are introduced in this work. The slip parameters are introduced such that they are non-decreasing by definition, i.e., $\dot{\lambda}_\alpha \geq 0$. Therefore, each slip parameter represents the occurred plastic slip into its corresponding slip direction. The plastic slip γ_α of a slip system is given by the difference of the two (positive) slip parameters $\lambda_\alpha - \lambda_{\alpha+12}$, corresponding to the positive and negative slip directions on slip system α. For an FCC-crystal this results in $\alpha = 1, \ldots, 24$ plastic slip parameters. The plastic strain tensor in terms of the plastic slip parameters is given by

$$\boldsymbol{\varepsilon}^{\mathrm{p}} = \sum_{\alpha} \lambda_\alpha \boldsymbol{M}^s_\alpha, \tag{2.5}$$

and the accumulated plastic slip in terms of the plastic slip parameters reads (Wulfinghoff and Böhlke, 2012; Wulfinghoff et al., 2013)

$$\gamma_{\mathrm{ac}}(\hat{\lambda}) = \sum_{\alpha} \int \dot{\lambda}_\alpha \, \mathrm{d}t = \sum_{\alpha} \lambda_\alpha. \tag{2.6}$$

In the following derivations, an additional so-called micromorphic field variable is used. It is denoted by ζ and is introduced due to numerical benefits regarding the implementation of the theory. The micromorphic variable is the (micro) counterpart to the (macro) field γ_{ac}. Due to the introduction of ζ, an additional stress quantity is present in the equations that follow. The two formulations, i.e., either solely in terms of γ_{ac} or in terms of both $\gamma_{\mathrm{ac}}, \zeta$, are equivalent. In order to show equivalence, an illustrative example of such an extension is discussed, at first.

2.2 Illustrative example for a micromorphic extension

The following example is based on an idea by Wulfinghoff (2011). A linear spring system is considered, loaded by a single force F (see Fig. 2.1a). This system has one degree of freedom (DOF), the displacement u.

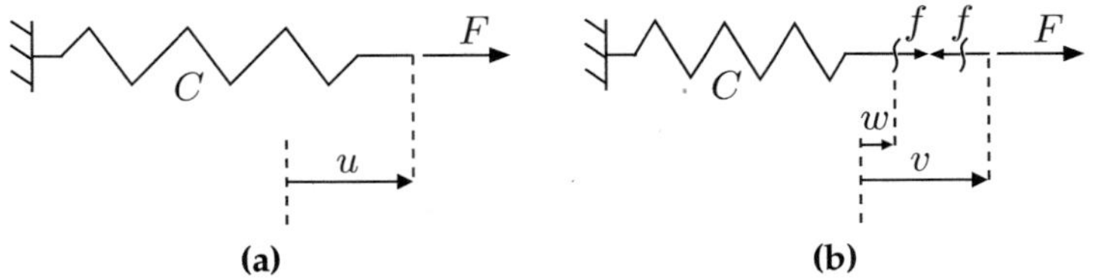

Figure 2.1: (a) Spring system with an imposed force F and one degree of freedom u. (b) Spring system with an imposed force F and three degrees of freedom w, v, f. Sketch based on an idea by Wulfinghoff (2011).

For a conservative force F, the potential energy of the depicted system (a) is given by

$$U(u) = \frac{1}{2}Cu^2 - Fu, \tag{2.7}$$

where the first term is the potential of the spring, and the second term is a potential of the force F. Variation of the potential energy with respect to the degree of freedom u gives

$$\delta U(u, \delta u) = (Cu - F)\delta u. \tag{2.8}$$

By the principle of stationary action, the displacement in the equilibrium state can be obtained via

$$\delta U(u, \delta u) = 0 \, \forall \delta u \Rightarrow u = \frac{F}{C}. \tag{2.9}$$

Additionally, a second system is considered with three DOFs: two displacements w, v, and an additional force f (see Fig. 2.1b). The potential energy of system (b) is given by

$$\tilde{U}(v, w, f) = \frac{1}{2}Cw^2 - Fv + f(v - w). \tag{2.10}$$

The variation of $\tilde{U}$ with respect to its DOFs reads

$$\delta\tilde{U}(v, w, f, \delta v, \delta w, \delta f) = (Cw - f)\delta w + (f - F)\delta v + (v - w)\delta f. \tag{2.11}$$

In the equilibrium state,

$$\delta\tilde{U}(v, w, f, \delta v, \delta w, \delta f) = 0 \, \forall \delta w, \delta v, \delta f, \tag{2.12}$$

and, thus, the resulting field equations are given by

$$Cw - f = 0, \tag{2.13}$$

$$f - F = 0, \tag{2.14}$$

$$v - w = 0. \tag{2.15}$$

The equivalence of both system formulations Fig. 2.1a, Fig. 2.1b can be shown by requiring $v = w = u$. For Eqs. (2.13–2.15), it follows that $u = F/C$.

It should be noted that the difference between the (original) DOF $u = v$ and the (additional) DOF w enters the potential energy, Eq. (2.10). However, if equality between both DOFs is required, the potential Eq. (2.10) reduces to the potential Eq. (2.7). Therefore, a minimization of the potential energy Eq. (2.10) is equivalent to a minimization of the potential energy Eq. (2.7), if the additional DOF w is required to be equal to its counterpart u. In a similar fashion as in this illustrative example, an additional field variable ζ is introduced as a micromorphic counterpart to the accumulated plastic slip γ_{ac} in the crystal plasticity framework of this work. Details are given in the next section.

2.3 Principle of virtual power, resulting field equations and boundary conditions

The field equations are derived using the principle of virtual power (POVP), see, e.g., Gurtin (2008). It states that the virtual power of the internal forces $\delta\mathcal{P}_{\mathrm{int}}$ equals the virtual power of the external forces $\delta\mathcal{P}_{\mathrm{ext}}$,

$$\delta\mathcal{P}_{\mathrm{int}} = \delta\mathcal{P}_{\mathrm{ext}}. \tag{2.16}$$

An internal power density of the bulk has been presented, for the single slip case, in Eq. (1.23). In the following, the internal power density is formulated for the multi-slip case. Contrary to physically more sophisticated crystal-plasticity theories (e.g., Gurtin, 2008), the plastic slips of all slip systems are not modeled as individual degrees of freedom. Instead, the accumulated plastic slip γ_{ac} is considered in the formulation, and a micromorphic variable ζ is introduced as an additional DOF. The internal power density of the bulk material, p_{vol}, is assumed to be given by the form

$$p_{\mathrm{vol}} = \boldsymbol{\sigma} \cdot \dot{\boldsymbol{\varepsilon}} + \pi\dot{\zeta}. \tag{2.17}$$

Formally, this statement is an extension of the classic power of the internal forces. An additional, generalized stress π is considered, which is work-conjugate to the rate $\dot{\zeta}$. In the absence of interfaces such as grain boundaries, the internal power of the bulk, $\mathcal{P}_{\mathrm{int}}$, reads

$$\mathcal{P}_{\mathrm{int}} = \int_{\mathcal{B}} p_{\mathrm{vol}} \,\mathrm{d}v. \tag{2.18}$$

The external power $\mathcal{P}_{\text{ext}}$ is assumed to be expanded on the external boundaries $\partial\mathcal{B}$ in the form

$$\mathcal{P}_{\text{ext}} = \int_{\partial\mathcal{B}} \bar{\boldsymbol{t}} \cdot \dot{\boldsymbol{u}} \, \mathrm{d}a, \tag{2.19}$$

where $\bar{\boldsymbol{t}}$ is a prescribed traction vector expending power via the rate $\dot{\boldsymbol{u}}$ of the displacement. In the following, $\dot{\boldsymbol{u}} = \delta\dot{\boldsymbol{u}}$ and $\dot{\zeta} = \delta\dot{\zeta}$ are virtual rates, vanishing at Dirichlet boundaries $\partial\mathcal{B}_{\mathrm{u}}$ for given $\{\boldsymbol{u}, \zeta\}$. Using the virtual rates, and combining Eqs. (2.16–2.19) gives

$$\int_{\mathcal{B}} \left(\boldsymbol{\sigma} \cdot \delta\dot{\boldsymbol{\varepsilon}} + \pi\delta\dot{\zeta}\right) \mathrm{d}v = \int_{\partial\mathcal{B}_{\mathrm{t}}} \bar{\boldsymbol{t}} \cdot \delta\dot{\boldsymbol{u}} \, \mathrm{d}a. \tag{2.20}$$

Here, it should be noted that the variables $\boldsymbol{\varepsilon}^{\mathrm{p}}$ and ζ are, a priori, chosen to be independent, i.e., $\delta\dot{\boldsymbol{\varepsilon}}^{\mathrm{p}} = 0$ (cf. Wulfinghoff, 2014).
Considering $\boldsymbol{\varepsilon} = \mathrm{sym}(\mathrm{grad}\,(\boldsymbol{u}))$ and applying form (B.1) of the divergence theorem to the first term in Eq. (2.20) leads, after regrouping of terms, to

$$-\int_{\mathcal{B}} \mathrm{div}\,(\boldsymbol{\sigma}) \cdot \delta\dot{\boldsymbol{u}} \, \mathrm{d}v + \int_{\partial\mathcal{B}_{\mathrm{t}}} (\boldsymbol{\sigma}\boldsymbol{n} - \bar{\boldsymbol{t}}) \cdot \delta\dot{\boldsymbol{u}} \, \mathrm{d}a + \int_{\mathcal{B}} \pi\delta\dot{\zeta} \, \mathrm{d}v = 0. \tag{2.21}$$

Since the principle of virtual power is valid for arbitrary virtual rates $\delta\dot{\boldsymbol{u}}$ and $\delta\dot{\zeta}$, the left-hand side of Eq. (2.21) has to vanish, independently of the virtual rates. Therefore, the set of field equations and boundary conditions (BCs) in Table 2.1 is obtained. Here, the classic linear momentum balance is extended by an additional so-called microforce balance in which the generalized stress π is considered. In the case that gradient contributions are neglected, no additional terms are present. If, however, gradient contributions like, for example, $\nabla\zeta$ are considered, an additional vectorial microstress is present in the microforce balance (see also Table 3.1).

Table 2.1: Field equations and boundary conditions.

Linear momentum balance	$\operatorname{div}(\boldsymbol{\sigma})$	$= \mathbf{0}$	$\forall \boldsymbol{x} \in \mathcal{B}$
Microforce balance	π	$= 0$	$\forall \boldsymbol{x} \in \mathcal{B}$
Cauchy stress Neumann BCs	$\boldsymbol{\sigma n}$	$= \bar{\boldsymbol{t}}$	on $\partial \mathcal{B}_t$

It should be noted that the term "microforce balance" is used in accordance with the notions of Gurtin (e.g., Gurtin, 2008). The form of the equation resembles a balance but the equation is not a balance equation in the classical sense.

2.4 Constitutive equations

A form for the free energy density may be assumed to be given by the additive relation

$$W(\boldsymbol{\varepsilon}, \hat{\lambda}) = W_{\mathrm{e}}(\boldsymbol{\varepsilon}, \boldsymbol{\varepsilon}^{\mathrm{p}}(\hat{\lambda})) + W_{\mathrm{h}}(\gamma_{\mathrm{ac}}(\hat{\lambda})), \tag{2.22}$$

where besides the classic, elastic energy density an additional contribution in form of a potential W_{h} for the hardening stress is assumed. It should be noted that it is a convenient choice to assume a potential for the hardening stress, from which the hardening stress can be derived. Instead, one could neglect this contribution, here, and postulate a hardening stress in the flow rule, subsequently. An extension of the free energy Eq. (2.22), considering the micromorphic variable ζ, can be formulated by

$$W(\boldsymbol{\varepsilon}, \hat{\lambda}, \zeta) = W_{\mathrm{e}}(\boldsymbol{\varepsilon}, \boldsymbol{\varepsilon}^{\mathrm{p}}(\hat{\lambda})) + W_{\mathrm{h}}(\zeta) + W_{\chi}(\zeta - \gamma_{\mathrm{ac}}(\hat{\lambda})). \tag{2.23}$$

The additional contribution W_{χ} is a so-called penalty energy density of numerical nature. It ensures that the micromorphic variable ζ is approximately equal to the accumulated plastic slip γ_{ac}.

In detail, the assumed three contributions to the free energy density are given by the following forms

$$W_{\mathrm{e}}(\boldsymbol{\varepsilon}, \boldsymbol{\varepsilon}^{\mathrm{p}}(\hat{\lambda})) = \frac{1}{2}(\boldsymbol{\varepsilon} - \boldsymbol{\varepsilon}^{\mathrm{p}}(\hat{\lambda})) \cdot \mathbb{C}[\boldsymbol{\varepsilon} - \boldsymbol{\varepsilon}^{\mathrm{p}}(\hat{\lambda})], \tag{2.24}$$

$$W_{\mathrm{h}}(\zeta) = (\tau_{\infty}^{\mathrm{C}} - \tau_0^{\mathrm{C}})\zeta + \frac{1}{\Theta}(\tau_{\infty}^{\mathrm{C}} - \tau_0^{\mathrm{C}})^2 \exp\left(-\frac{\Theta\zeta}{\tau_{\infty}^{\mathrm{C}} - \tau_0^{\mathrm{C}}}\right), \tag{2.25}$$

$$W_{\chi}(\zeta - \gamma_{\mathrm{ac}}(\hat{\lambda})) = \frac{1}{2} H_{\chi}(\zeta - \gamma_{\mathrm{ac}}(\hat{\lambda}))^2. \tag{2.26}$$

The parameter H_χ is a numerical penalty-parameter. For a sufficiently large choice of H_χ, the micromorphic variable is $\zeta \approx \gamma_{\mathrm{ac}}$. In the limit case of $H_\chi \to \infty \Rightarrow \zeta = \gamma_{\mathrm{ac}}$.

For a purely mechanical theory, the total dissipation reads

$$D_{\mathrm{tot}} = \mathcal{P}_{\mathrm{ext}} - \int_{\mathcal{B}} \dot{W} \,\mathrm{d}v, \tag{2.27}$$

with the power of the external forces $\mathcal{P}_{\mathrm{ext}}$ being equal to the power of the internal forces $\mathcal{P}_{\mathrm{int}}$ (see also Eq. (2.18))

$$\mathcal{P}_{\mathrm{ext}} = \mathcal{P}_{\mathrm{int}} = \int_{\mathcal{B}} \boldsymbol{\sigma} \cdot \dot{\boldsymbol{\varepsilon}} + \pi \dot{\zeta} \,\mathrm{d}v. \tag{2.28}$$

Thus, the total dissipation is given by the difference between the power of the external forces (or the power of the internal forces, respectively) and the rate of the free energy. Considering the bulk dissipation as a volume integral over a dissipation density $\mathcal{D}$,

$$D_{\mathrm{tot}} = \int_{\mathcal{B}} \mathcal{D} \,\mathrm{d}v, \tag{2.29}$$

this dissipation density $\mathcal{D}$ can be specified. The combination of Eq. (2.23) and Eqs. (2.27, 2.28, 2.29) leads to

$$\mathcal{D} = \left(\boldsymbol{\sigma} - \frac{\partial W_{\mathrm{e}}}{\partial \boldsymbol{\varepsilon}}\right) \cdot \dot{\boldsymbol{\varepsilon}} - \frac{\partial W_{\mathrm{e}}}{\partial \boldsymbol{\varepsilon}^{\mathrm{p}}} \cdot \dot{\boldsymbol{\varepsilon}}^{\mathrm{p}} + \left(\pi - \frac{\partial W_{\mathrm{h}}}{\partial \zeta} - \frac{\partial W_{\chi}}{\partial \zeta}\right) \dot{\zeta} - \frac{\partial W_{\chi}}{\partial \gamma_{\mathrm{ac}}} \dot{\gamma}_{\mathrm{ac}} \geq 0. \tag{2.30}$$

In the following, it is assumed that $\boldsymbol{\sigma}$ and π are purely energetic stresses, i.e., they can be derived from a potential. Then, the two equalities

$$\boldsymbol{\sigma} = \frac{\partial W_{\mathrm{e}}}{\partial \boldsymbol{\varepsilon}}, \tag{2.31}$$

$$\pi = \frac{\partial W_{\mathrm{h}}}{\partial \zeta} + \frac{\partial W_{\chi}}{\partial \zeta} \tag{2.32}$$

can be deduced from Eq. (2.30). Therefore, the reduced dissipation inequality reads

$$\mathcal{D} = -\frac{\partial W_{\mathrm{e}}}{\partial \boldsymbol{\varepsilon}^{\mathrm{p}}} \cdot \dot{\boldsymbol{\varepsilon}}^{\mathrm{p}} - \frac{\partial W_{\chi}}{\partial \gamma_{\mathrm{ac}}} \dot{\gamma}_{\mathrm{ac}} \geq 0. \tag{2.33}$$

It is noted that $-\partial W_{\mathrm{e}}/\partial \boldsymbol{\varepsilon}^{\mathrm{p}} = \boldsymbol{\sigma}$, and that $\partial W_{\chi}/\partial \gamma_{\mathrm{ac}} = -\partial W_{\chi}/\partial \zeta$. The latter quantity is abbreviated by $\check{p}$, in the following. This is, in fact, an additional stress which arises due to the micromorphic approach. In detail, it reads

$$\check{p} = -\frac{\partial W_{\chi}}{\partial \zeta} = \frac{\partial W_{\chi}}{\partial \gamma_{\mathrm{ac}}} = H_{\chi}\left(\gamma_{\mathrm{ac}}(\hat{\lambda}) - \zeta\right). \tag{2.34}$$

Here, it becomes apparent that for $\zeta \to \gamma_{\mathrm{ac}} \Rightarrow \check{p} \to 0$.
The reduced dissipation inequality, considering the above introduced abbreviation, is given by

$$\mathcal{D} = \boldsymbol{\sigma} \cdot \dot{\boldsymbol{\varepsilon}}^{\mathrm{p}} - \check{p}\, \dot{\gamma}_{\mathrm{ac}} \geq 0. \tag{2.35}$$

Reformulating Eq. (2.35) in terms of the rates of the plastic slip parameters λ_α by exploiting Eq. (2.3) and Eq. (2.6) leads to

$$\mathcal{D} = \sum_\alpha (\boldsymbol{\sigma} \cdot \boldsymbol{M}^s_\alpha - \check{p})\, \dot{\lambda}_\alpha \geq 0 \tag{2.36}$$

with the symmetric part of the Schmid tensor $\boldsymbol{M}^s_\alpha = \mathrm{sym}(\boldsymbol{d}_\alpha \otimes \boldsymbol{n}_\alpha)$. The scalar product of the Cauchy stress and the symmetric part of the Schmid tensor gives the resolved shear stresses $\tau_\alpha = \boldsymbol{\sigma} \cdot \boldsymbol{M}^s_\alpha$. It is assumed that the bulk material dissipation is induced by the superposition of dissipative shear stresses τ^{d}_α from the individual slip systems (e.g., Cermelli and Gurtin, 2002)

$$\mathcal{D} = \sum_\alpha \tau^{\mathrm{d}}_\alpha \dot{\lambda}_\alpha. \tag{2.37}$$

From Eq. (2.36) and Eq. (2.37) it can be concluded that

$$\tau^{\mathrm{d}}_\alpha = \tau_\alpha - \check{p}. \tag{2.38}$$

Furthermore, using Eq. (2.34) in Eq. (2.32), the following relation is obtained for the numerical stress $\check{p}$

$$\check{p} = -\pi + \frac{\partial W_{\mathrm{h}}}{\partial \zeta}, \tag{2.39}$$

where the second term on the right-hand side is a hardening stress that is abbreviated by $\beta = \partial W_{\mathrm{h}} / \partial \zeta$, in the following. Combining Eq. (2.38) and Eq. (2.39) leads to dissipative shear stresses in form of

$$\tau^{\mathrm{d}}_\alpha = \tau_\alpha + \pi - \beta. \tag{2.40}$$

Due to the microforce balance (Table 2.1), $\pi = 0$, and the dissipative shear stresses reduce to

$$\tau^{\mathrm{d}}_\alpha = \tau_\alpha - \beta. \tag{2.41}$$

The flow rule for the bulk material is assumed to be of overstress type, formulated in the plastic slip parameter rates $\dot{\lambda}_\alpha$,

$$\dot{\lambda}_\alpha = \dot{\gamma}_0 \left\langle \frac{\tau_\alpha^{\mathrm{d}} - \tau_0^{\mathrm{C}}}{\tau^{\mathrm{D}}} \right\rangle^p , \tag{2.42}$$

and, by means of Eq. (2.41),

$$\dot{\lambda}_\alpha = \dot{\gamma}_0 \left\langle \frac{\tau_\alpha - (\tau_0^{\mathrm{C}} + \beta)}{\tau^{\mathrm{D}}} \right\rangle^p . \tag{2.43}$$

The finite element implementation of the presented theory in this chapter is a special case of the more general implementation which considers gradients of the micromorphic variable as well as GB contributions to the free energy. For brevity, the implementation of the special case at hand is not described in detail. The particulars of the more general implementation are given in Section 3.4.

2.5 Qualitative assessment of accumulated plastic slip distribution

2.5.1 Motivation

The simplification in the modeling of using the accumulated plastic slip γ_{ac}, instead of considering the individual slips γ_α of all slip systems as DOFs, raises the question whether this significant reduction in the DOFs is still reasonable regarding the physicality of the numerical results. Due to a lack of in-house availability of an implementation of a more sophisticated crystal-plasticity (CP) model, a direct comparison with a physically richer model could not be performed within the scope of the present work. Instead, a qualitative assessment of the model from Chapter 2 in comparison to experimental results of torsion-deformed pseudo-single-crystalline gold microwires is carried out. Details of the experimental approach are given in the collaborative publication by Ziemann et al. (2015). The experimental characterization and evaluation was performed by M. Ziemann, M. Walter and P.A. Gruber from the Institute of Applied Materials, KIT. Here, a brief summary of the experiments is given and it is focused on the comparison with CP results. Large parts of the CP results and discussion are taken from the modeling contribution to the joint work Ziemann et al. (2015).

2.5.2 Summary of the experimental characterization and torsion test results

Polycrystalline microwires were produced from high-purity gold by long-time annealing at high temperatures by Ziemann et al. (2015). After the annealing process, the cross-sections of the wires are predominantly single-crystalline. Thereby, a so-called bamboo-like structure of the wires is achieved, i.e., the wires consist of a sequence of single-crystalline grains along their central axes. Different orientations of the grains are observed. By electronic backscatter diffraction, grains of $\langle 100 \rangle$-crystal-orientation (with respect to the central axis) are identified. Grains are chosen which are large enough (in lateral direction) to allow for an evaluation of their cross-sections unimpeded by GB influences. By using the focused ion beam technique, cross-sections are extracted at the center of the grains. Subsequently, the cross-sections are characterized by Laue-microdiffraction experiments. The obtained diffraction data allows to calculate the misorientation of the cross-sections for each data point with respect to the central fiber. By this procedure, a measure of global deformation is obtained. Additionally, the local deformation is calculated by means of the kernel average misorientation (KAM), which represents the averaged misorientation of a data point, obtained by averaging over the misorientations of the eight nearest (neighboring) data points.Results of the normalized (global) misorientation maps are depicted for cross-sections obtained from different torsion-deformed microwires at several maximum plastic shear values $\gamma_{r=R}$ in Fig. 2.2.

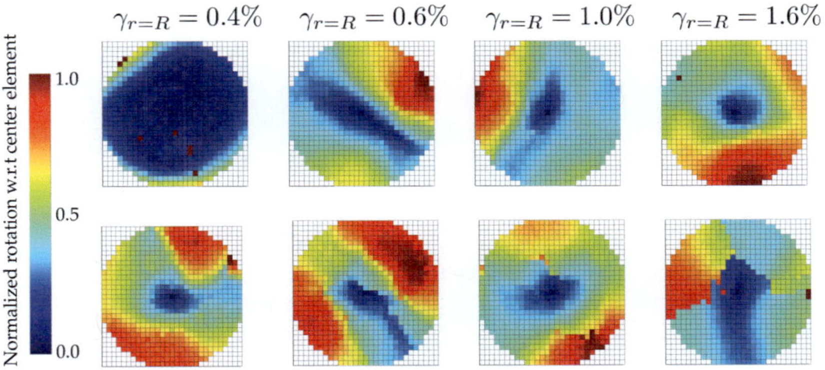

Figure 2.2: Normalized misorientation maps from cross-sections of different twisted bamboo-structured Au microwires with a diameter of 25 µm. The corresponding grains have orientations close to the [001]-orientation with respect to the experimental (out-of-plane) rotation axis. In this analysis, the rotation of every data point is referred to the center element and the misorientation is normalized to the maximum misorientation that occurred in the particular cross-section. Figure reprinted from Ziemann et al. (2015) with permission from Elsevier. Images courtesy of M. Ziemann.

It is observable that pronounced differences are present in the global deformation maps of the cross-sections of different wires. Some of the cross-sections show roughly a double-fold symmetry of the misorientation. For the same cross-sections, the misorientation is averaged over circular line segments and plotted over the distance to the center of the cross-sections in Fig. 2.3a (averaged for two different samples at each macroscopic plastic deformation). Considering the averaged misorientation distribution in radial direction, it is observable that an almost linear increase from the center towards the lateral surface is present for small overall plastic deformations.

With increasing overall plastic deformation, the misorientation development saturates into a plateau-like distribution.

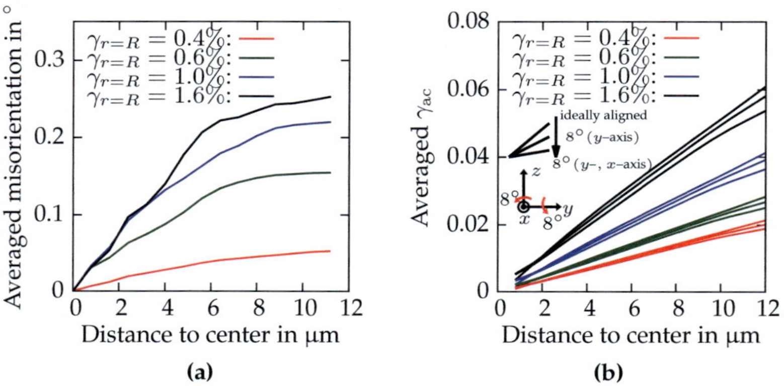

Figure 2.3: (a) Averaged misorientation vs. distance to the center, calculated for the cross-sections shown in Fig. 2.2, averaged for both grains extracted from the same wire. (b) Averaged accumulated plastic slip vs. distance to center, calculated for the cross-sections shown in Figs. 2.5–2.7. Figures reprinted from Ziemann et al. (2015) with permission from Elsevier. Experimental data courtesy of M. Ziemann.

Close to the surface of the microwires, the result, therefore, contradicts an idealized, solely linear plastic-deformation gradient in radial direction. It is possible that this is due the annihilation of a substantial part of the dislocations close to the surface (Ziemann et al., 2015).

Additionally, the KAM-distributions of the cross-sections are depicted in Fig. 2.4. Comparison of the visible deformation traces in Fig. 2.4 to the global deformation maps in Fig. 2.2 reveals that the traces are mainly correlating to the borders of regions that are substantially deformed (Ziemann et al., 2015). For the case of an $\langle 100 \rangle$-crystal-orientation ideally aligned with the wire axis, one would expect a fourfold symmetry of the global deformation field due to four slip systems that are equally favorable for an activation under the imposed torsion loading. However, the experimental results do not exhibit this type of plastic behavior. This can be explained by considering deviations from the ideal orientation that occurred in the preparation of the experimental samples (Ziemann et al., 2015).

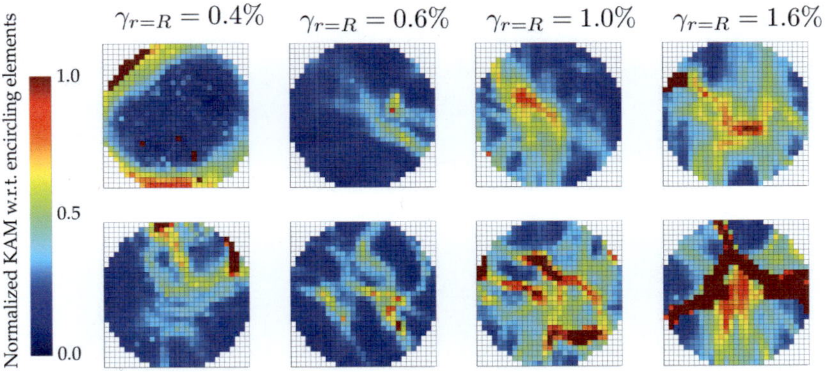

Figure 2.4: Normalized kernel average misorientation maps from cross-sections of different twisted bamboo-structured Au microwires with a diameter of 25 µm. The arrows show the in-plane ⟨100⟩-orientations. Figure reprinted from Ziemann et al. (2015) with permission from Elsevier. Images courtesy of M. Ziemann.

Exemplary, this is investigated using a finite element implementation of the preceding crystal plasticity model, in the following.

2.5.3 Crystal plasticity simulation results

In order to qualitatively compare a measure of deformation from the simulations to the experiments, crystal plasticity simulations with γ_{ac} as (plastic) deformation measure are carried out within a micromorphic implementation. Thereby, field distributions of the accumulated plastic slip are obtained. It is remarked that the use of the micromorphic approximation of γ_{ac} in the simulations is feasible if the penalty parameter H_χ is chosen large enough and a sufficiently fine finite element mesh is used (see Appendix C.1 for a comparison of both field distributions

for the single-crystalline case of this section).
A cylindrical simulation volume of diameter 25 μm and of height 31.25 μm is considered, at first, in the ideally aligned ⟨100⟩-orientation. The top surface and the bottom surface are rotated, relatively to each other, until a relative maximum plastic shear of $\gamma_{r=R} = 0.02$ is reached. In lateral direction, the displacement of the top and bottom nodes is unrestricted, except for one node that is spatially fixed (for uniqueness of the position in space). The model parameters are listed in Table 2.2.

Table 2.2: Model parameters for simulations of gold single-crystals.

C_{1111}	C_{1122}	C_{1212}	H_χ	p
168 GPa	121 GPa	75 GPa	10^6 MPa	20
$\dot{\gamma}_0$	Θ	τ_0^{C}	τ_∞^{C}	τ^{D}
10^{-3} 1/s	330 MPa	6 MPa	55 MPa	1 MPa

In Fig. 2.5, the resulting distributions of the accumulated plastic slip are depicted, at the same fixed values of the plastic overall shear as in the experiments. As it is expected for the ideally with the central axis aligned ⟨100⟩-crystal-orientation, a fourfold symmetry of the accumulated plastic slip distribution is obtained and evolves with the overall plastic deformation. In addition, simulations are carried out that consider deviations from the ideal alignment of the crystal orientation with respect to the rotation axis. As also discussed above, these deviations occur in the experimental setup due to the imperfect production of ideally with the central axis aligned crystal-orientations as well as small misalignments in the experimental setup (Ziemann et al., 2015).

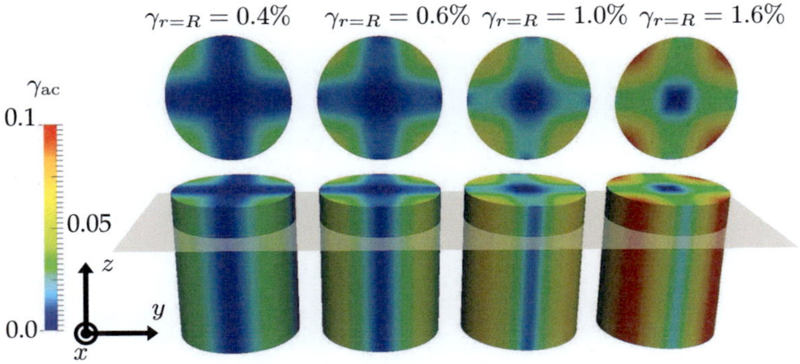

Figure 2.5: CP simulation: Evolution of the spatial distribution of the accumulated plastic slip with increasing twist within an ideally aligned ⟨100⟩-oriented single-crystalline cylinder of Au. Figure reprinted from Ziemann et al. (2015) with permission from Elsevier.

In Fig. 2.6, the accumulated plastic slip distribution is shown for a misalignment of 8° around one in-plane axis, and, in Fig. 2.7, for the case of an additional misalignment of 8° around the second in-plane axis. The corresponding overall mechanical responses are plotted in Fig. 2.9b, showing a comparably small effect of the misorientations.

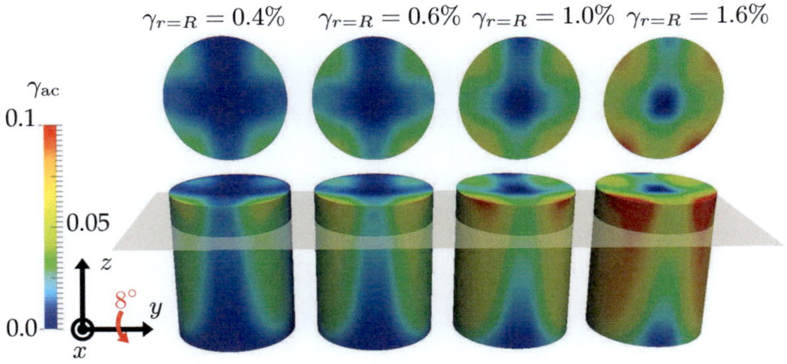

Figure 2.6: CP simulation: Evolution of the spatial distribution of the accumulated plastic slip with increasing twist within a misaligned ⟨100⟩-oriented single-crystalline cylinder of Au. Misalignment of 8° related to the y-axis. Figure reprinted from Ziemann et al. (2015) with permission from Elsevier.

However, it is observable that significant changes in the distributions of γ_{ac} are caused by the misorientations (Figs. 2.6–2.7). In both cases, the fourfold-symmetry distribution changes to different twofold symmetric distributions, also clearly visible in the misorientation distribution, see Fig. 2.2 ($\gamma_{r=R} = 0.6\%$, bottom image).

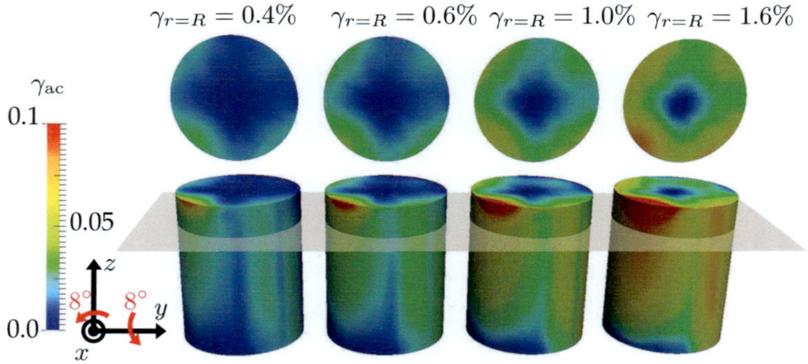

Figure 2.7: CP simulation: Evolution of the spatial distribution of the accumulated plastic slip with increasing twist within a misaligned ⟨100⟩-oriented single-crystalline cylinder of Au. Misalignment of 8° related to both the x-axis and the y-axis. Figure reprinted from Ziemann et al. (2015) with permission from Elsevier.

Additionally, in the experimental KAM-maps, high-deformation areas are observed that could correlate to regions of large gradients of γ_{ac}. The in-plane gradient components of γ_{ac} are depicted for the different cases in Fig. 2.8. High gradients of γ_{ac} are observable in between regions of significantly differing plastic deformations. The symmetry of the gradient distributions is affected by the misorientation (see Figs. 2.8d–2.8i). The gradient distributions in Figs. 2.8b–2.8c show similar maxima as, for example, the KAM-map in Fig. 2.4 (top image for $\gamma_{r=R} = 1.6\%$). Consequently, results qualitatively, at least partly, similar to the experimental data are obtained by the modeling approach with γ_{ac}.

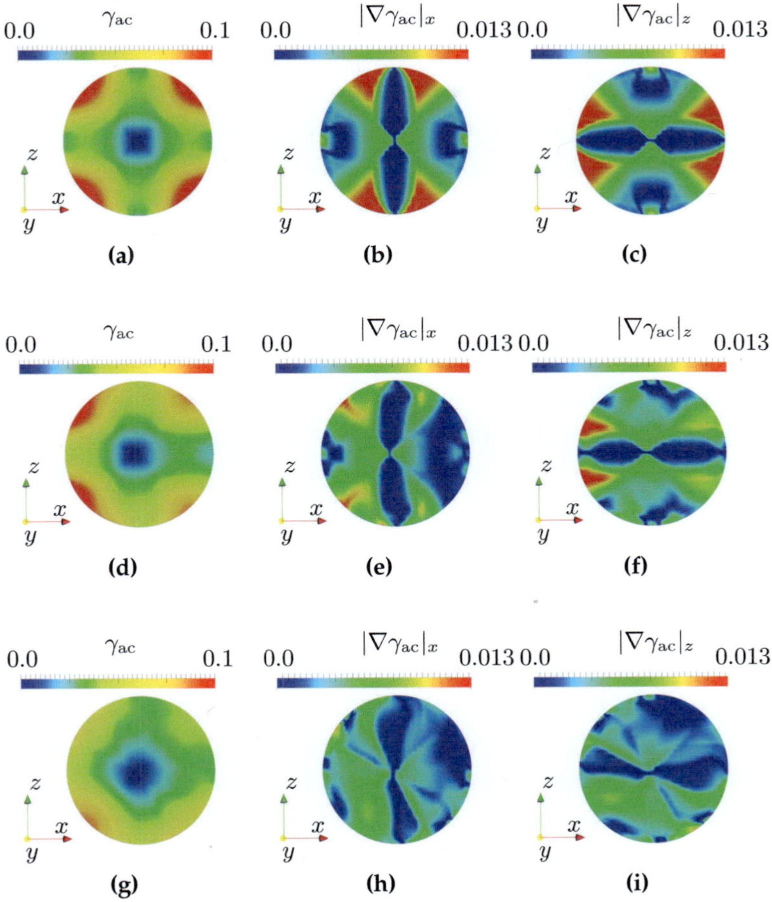

Figure 2.8: (a) Field distribution of γ_{ac} in the cross-section indicated in Fig. 2.5, (b) corresponding field distribution of $|\nabla\gamma_{ac}|_x$, and (c) of $|\nabla\gamma_{ac}|_z$. (d) Field distribution of γ_{ac} in the cross-section indicated in Fig. 2.6, (e) corresponding field distribution of $|\nabla\gamma_{ac}|_x$, and (f) of $|\nabla\gamma_{ac}|_z$. (g) Field distribution of γ_{ac} in the cross-section indicated in Fig. 2.7, (h) corresponding field distribution of $|\nabla\gamma_{ac}|_x$, and (i) of $|\nabla\gamma_{ac}|_z$.

It is remarked that only a first comparison has been performed and a quantitative assessment was not performed. This would require to calculate the misorientation of the numerical results. For a direct correlation of simulation and experimental results, also, many more cross-sections would need to be characterized, than they could be evaluated within the scope of Ziemann et al. (2015). Otherwise, the influence by dislocation substructures (Ziemann et al., 2015) on the experimental results cannot be entirely ruled out as a possible influence. However, qualitative correspondence of the model results to the experimental data is demonstrated, at least to some extent, considering the radial distributions of γ_{ac}, averaged over circular line segments (Fig. 2.3b). With evolving plastic deformation, an overall increase in the gradients of this measure, similar to the increase in the experimental misorientation-gradients, Fig. 2.3a, is observable. As visible in Fig. 2.3b, the small misorientations, however, lead only to moderate differences in the radial distributions of γ_{ac}. Therefore, these misorientations cannot explain the substantial differences occurring for different experimental samples which have been shown in Ziemann et al. (2015). In addition, the γ_{ac}-distributions show an almost ideal, linear course. This is the case because annihilation of dislocations is not accounted for in the theory, due to the non-decreasing plastic slip parameters.

It is suggested to, additionally, investigate also the interaction mechanisms of dislocations across GBs between smaller grains of different orientations in further experimental works and accompanying crystal plasticity simulations, in the future. For this, a physically more sophisticated model should be used that allows to also reproduce the annihilation of dislocations close to the surface.

2.5.4 Influence of $\langle 100 \rangle$- and $\langle 111 \rangle$-orientations on the mechanical response under tensile and torsion loading

One of the goals of this work is the modeling of size effects under different loading conditions such as tensile and torsion loading (e.g., in the work by Chen et al., 2015). In a collaborative effort, microwires of different diameters are investigated both with experimental characterization and numerical simulations by Bayerschen et al. (2016b). In the simulations, gradient contributions are considered. The corresponding results are discussed in Section 3.6. In the investigations, a non-negligible influence by the crystal orientation of different grains on the mechanical response is found. For some of the investigated microwires, mainly two orientations are present in the experimentally characterized cross-sections. Therefore, in the current section of this thesis, the influence of these two crystal orientations is investigated, isolated from the gradient- and GB contributions, at first. The computational implementation of the preceding mathematical framework, from Chapter 2, is exploited, therefore. Subsequently, in Section 3.6.6, the discussion is refined taking the gradient and GB contributions into account.

Simulations of the above-described single-crystalline cylindrical volume of diameter $25\,\mu\text{m}$ are carried out for two crystal-orientations. Torsion test simulation results are compared to tensile test simulation results: at first, for the above-investigated $\langle 100 \rangle$-orientation aligned with the central wire-axis, then, additionally, for the case that the $\langle 111 \rangle$-orientation is aligned with the central wire-axis.

In Fig. 2.9, for both crystal orientations, the tensile stress-strain curves are compared to the maximum shear stresses $\tau_{r=R} = 2M_T/(\Pi R^3)$ (M_T: torque , R: max. radius) over the maximum shear for torsion loading. It is observable that the $\langle 100 \rangle$-orientation is behaving substantially softer under tensile loading than the $\langle 111 \rangle$-orientation (Fig. 2.9a).

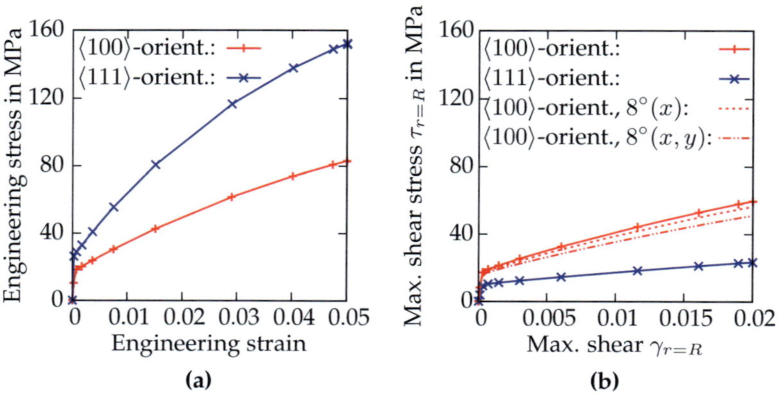

Figure 2.9: (a) Stress-strain curves of two single-crystals with different orientations under tensile loading. (b) Maximum shear stress vs. maximum shear of the single-crystals from (a) under torsion loading. Additionally shown are the torsion-responses of the single-crystals for misorientations from the ideally aligned orientation, see Figs. 2.6–2.7.

Under torsion loading, this effect is reversed (Fig. 2.9b). Since gradient and GB contributions are neglected, this effect can only stem from the employed crystal orientations. The pronounced differences in the orientations of the slip systems with regard to the loading direction for the different crystal orientations cause the differences in the overall mechanical responses. In Fig. 2.10, the plastic slip contributions to γ_{ac} are depicted for the employed torsion loading. The used slip system convention can be found in Appendix E.1. For the considered case, only three slip systems are active ($\alpha = 7 - 9$, see Appendix E.2 for their spatial orientation). These slip systems are located on the $\langle 111 \rangle$-slip plane (see Fig. 1.5b) which is perfectly aligned with the lateral cross-sections of the simulation volume, here. Therefore, these systems are favorable for an activation of plastic slip. In contrast, for the $\langle 100 \rangle$-orientation, mainly four slip systems ($\alpha = 1, 4, 7, 10$) contribute to γ_{ac} under torsion loading (Fig. 2.11). These slip systems are oriented diagonally within the surface planes of the FCC unit-cell (see Fig. 1.5c).

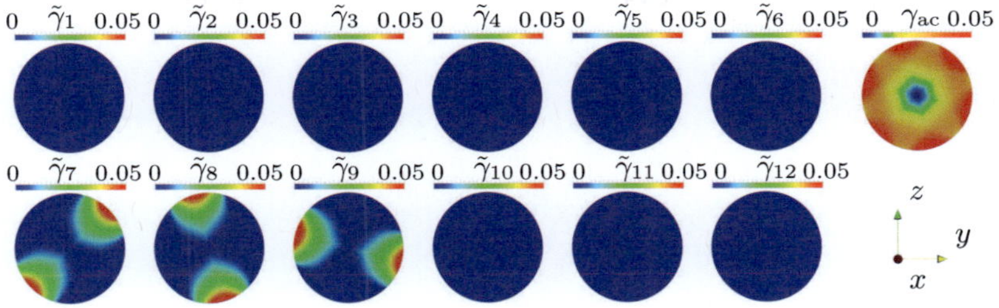

Figure 2.10: Accumulated plastic slip contributions $\tilde{\gamma}_\alpha = \lambda_\alpha + \lambda_{\alpha+12}$ of the individual slip systems $\alpha = 1, \ldots, 12$ to the accumulated plastic slip γ_{ac} for an $\langle 111 \rangle$-oriented crystal under torsion loading. The slip system convention is listed in Appendix E.1.

Since for the $\langle 100 \rangle$-orientation, these slip planes of the FCC unit-cell are perfectly aligned with the lateral cross-sections of the simulation volume, the corresponding slip systems are favorably oriented to be activated during the plastic deformation process. The activation of these slip systems causes the fourfold symmetry of γ_{ac} (Fig. 2.11).

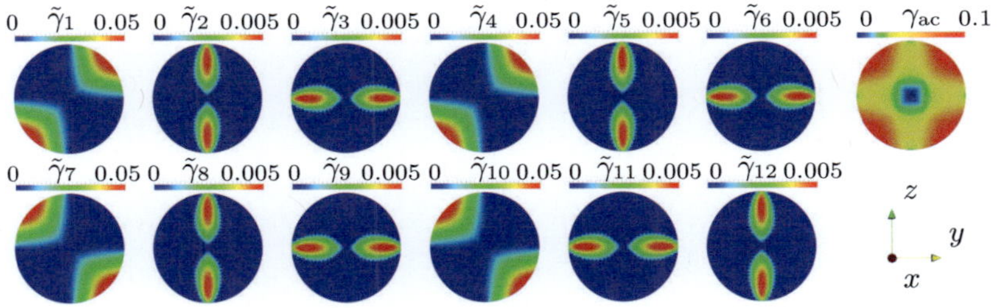

Figure 2.11: Accumulated plastic slip contributions $\tilde{\gamma}_\alpha = \lambda_\alpha + \lambda_{\alpha+12}$ of the individual slip systems $\alpha = 1, \ldots, 12$ to the accumulated plastic slip γ_{ac} for an $\langle 100 \rangle$-oriented crystal under torsion loading. The slip system convention is listed in Appendix E.1.

However, all other slip systems are also activated, leading to non-zero γ_{ac} values in the regions between the maximum γ_{ac} values. For the same applied maximum shear, the single crystal with $\langle 111 \rangle$-orientation shows a substantially smaller magnitude of γ_{ac}, compared to the one

with $\langle 100 \rangle$-orientation. Thus, the $\langle 100 \rangle$-orientation responds stiffer for this loading type (see Fig. 2.9b).
Concluding from the single-crystal results of this section, the influence of the crystal-orientations cannot a priori be neglected when investigating and simulating the mechanical response of oligocrystalline microwires with several grains in $\langle 111 \rangle$- and $\langle 100 \rangle$-orientation.

Chapter 3

Gradient crystal plasticity with an accumulated plastic slip and grain boundary yielding

3.1 Motivation

In microstructured non-single-crystalline materials, the presence of GBs leads to different mechanical responses, compared to single-crystalline materials. Therefore, this influence needs to be considered when modeling such materials on the continuum scale. The micromorphic accumulated plastic slip framework from the preceding chapter is extended to account for the influence of GBs that impede dislocation movement. Therefore, a GB yield condition is introduced. This condition provides a mechanism of "adjusting" the yield strength of the GBs, which then, in turn, phenomenologically models mechanisms such as transfer of dislocations and deposition of dislocations at the GBs. In addition, the principle of virtual power is enhanced to consider the arising contributions from the occurring gradients of the micromorphic variable ζ at the GBs and at the boundaries. Large parts of this chapter are based on the works Wulfinghoff et al. (2013) and Bayerschen et al. (2015).

3.2 Gradient-extended principle of virtual power, resulting field equations and boundary conditions

The basic assumptions from Section 2.1 are assumed to hold in the present chapter, too. By extending the internal power density of the bulk, Eq. (2.17), with a $\nabla\zeta$-contribution, the following is obtained

$$p_{\mathrm{vol}} = \boldsymbol{\sigma} \cdot \dot{\boldsymbol{\varepsilon}} + \pi\dot{\zeta} + \boldsymbol{\xi} \cdot \nabla\dot{\zeta}. \tag{3.1}$$

Here, $\boldsymbol{\xi}$ is a generalized or gradient-related vectorial stress, work-conjugate to $\nabla\dot{\zeta}$. In addition, it is assumed that the internal power of the bulk can be decomposed into a volumetric contribution and a GB contribution

$$\mathcal{P}_{\mathrm{int}} = \int_{\mathcal{B}} p_{\mathrm{vol}} \,\mathrm{d}v + \int_{\Gamma} \Xi_{\Gamma}\dot{\zeta} \,\mathrm{d}a, \tag{3.2}$$

where Ξ_{Γ} is a GB microtraction. This microtraction is imposed in form of a jump-condition for the generalized stress $\boldsymbol{\xi}$, as it is shown below. Possible jump contributions of ζ (at the GBs) are neglected, to keep the number of constitutive equations and necessary new material parameters low. Considering an extension of the external power $\mathcal{P}_{\mathrm{ext}}$ from Eq. (2.19), the following is postulated

$$\mathcal{P}_{\mathrm{ext}} = \int_{\partial\mathcal{B}} \left(\bar{\boldsymbol{t}} \cdot \dot{\boldsymbol{u}} + \bar{\Xi}\dot{\zeta}\right) \,\mathrm{d}a. \tag{3.3}$$

Here, the quantity $\bar{\Xi}$ is a prescribed microtraction, expending power via the rate $\dot{\zeta}$ of the micromorphic variable. The boundary $\partial\mathcal{B}$ can be divided into $\partial\mathcal{B} = \partial\mathcal{B}_{\mathrm{t}} \cup \partial\mathcal{B}_{\Xi} \cup \partial\mathcal{B}_{\mathrm{u}}$ with $\partial\mathcal{B}_{\mathrm{t}} \cup \partial\mathcal{B}_{\Xi} \cap \partial\mathcal{B}_{\mathrm{u}} = \varnothing$.

Furthermore, the gradient-extension of the POVP (Eq. (2.16)) reads

$$\int_{\mathcal{B}} \left(\boldsymbol{\sigma} \cdot \delta\dot{\boldsymbol{\varepsilon}} + \pi\delta\dot{\zeta} + \boldsymbol{\xi} \cdot \nabla\delta\dot{\zeta}\right) \mathrm{d}v + \int_{\Gamma} \Xi_{\Gamma}\delta\dot{\zeta}\,\mathrm{d}a = \int_{\partial\mathcal{B}_{\mathrm{t}}} \bar{\boldsymbol{t}} \cdot \delta\dot{\boldsymbol{u}}\,\mathrm{d}a + \int_{\partial\mathcal{B}_{\Xi}} \bar{\Xi}\,\delta\dot{\zeta}\,\mathrm{d}a. \tag{3.4}$$

Then, by substituting $\boldsymbol{\varepsilon} = \mathrm{sym}(\nabla\boldsymbol{u})$ in Eq. (3.4), taking into account the chain rule as well as $\boldsymbol{\sigma} = \boldsymbol{\sigma}^{\mathsf{T}}$, and by applying the divergence theorem in the form (B.1) to the first term of the volume integral in Eq. (3.4), the following form is obtained

$$\int_{\mathcal{B}} \boldsymbol{\sigma} \cdot \delta\dot{\boldsymbol{\varepsilon}}\,\mathrm{d}v = \int_{\mathcal{B}} -\mathrm{div}\,(\boldsymbol{\sigma}) \cdot \delta\dot{\boldsymbol{u}}\,\mathrm{d}v + \int_{\partial\mathcal{B}_{\mathrm{t}}} \boldsymbol{\sigma}\boldsymbol{n} \cdot \delta\dot{\boldsymbol{u}}\,\mathrm{d}a. \tag{3.5}$$

In (B.1), the continuity of $\boldsymbol{t} = \boldsymbol{\sigma}\boldsymbol{n}$ has been considered, already. Consequently, jump terms at the GB Γ are neglected. The application of the divergence theorem of the form (B.3) to the third term of the volume integral in Eq. (3.4) gives

$$\int_{\mathcal{B}} \boldsymbol{\xi} \cdot \nabla\delta\dot{\zeta}\,\mathrm{d}v = \int_{\mathcal{B}} -\mathrm{div}\,(\boldsymbol{\xi})\,\delta\dot{\zeta}\,\mathrm{d}v + \int_{\partial\mathcal{B}_{\Xi}} \boldsymbol{\xi} \cdot \boldsymbol{n}\,\delta\dot{\zeta}\,\mathrm{d}a - \int_{\Gamma} [\![\boldsymbol{\xi}]\!] \cdot \boldsymbol{n}\,\delta\dot{\zeta}\,\mathrm{d}a. \tag{3.6}$$

After substitution of Eqs. (3.5) and (3.6) into Eq. (3.4) and regrouping of terms, the following is obtained

$$\begin{aligned} -\int_{\mathcal{B}} \mathrm{div}\,(\boldsymbol{\sigma}) \cdot \delta\dot{\boldsymbol{u}}\,\mathrm{d}v + \int_{\mathcal{B}} (\pi - \mathrm{div}\,(\boldsymbol{\xi}))\,\delta\dot{\zeta}\,\mathrm{d}v + \int_{\partial\mathcal{B}_{\mathrm{t}}} (\boldsymbol{\sigma}\boldsymbol{n} - \bar{\boldsymbol{t}}) \cdot \delta\dot{\boldsymbol{u}}\,\mathrm{d}a \\ + \int_{\partial\mathcal{B}_{\Xi}} (\boldsymbol{\xi} \cdot \boldsymbol{n} - \bar{\Xi})\delta\dot{\zeta}\,\mathrm{d}a + \int_{\Gamma} (\Xi_{\Gamma} - [\![\boldsymbol{\xi}]\!] \cdot \boldsymbol{n})\,\delta\dot{\zeta}\,\mathrm{d}a = 0. \end{aligned} \tag{3.7}$$

Requiring the left-hand side of Eq. (3.7) to vanish for arbitrary $\delta\dot{\boldsymbol{u}}, \delta\dot{\zeta}$ yields the field equations and GB conditions / BCs in Table 3.1.

Table 3.1: Field equations and boundary conditions with gradient extension. The jump of $\boldsymbol{\xi}$ is denoted by $[\![\boldsymbol{\xi}]\!] = \boldsymbol{\xi}^+ - \boldsymbol{\xi}^-$. The GB normal points from "-" to "+". Table reprinted from Wulfinghoff et al. (2013) with permission from Elsevier.

Linear momentum balance	$\mathbf{0}$	$= \mathrm{div}\,(\boldsymbol{\sigma})$	$\forall \boldsymbol{x} \in \mathcal{B}$
Microforce balance	π	$= \mathrm{div}\,(\boldsymbol{\xi})$	$\forall \boldsymbol{x} \in \mathcal{B} \setminus \Gamma$
GB microtraction	Ξ_Γ	$= [\![\boldsymbol{\xi}]\!] \cdot \boldsymbol{n}$	$\forall \boldsymbol{x} \in \Gamma$
Neumann BCs for: Cauchy stress	$\boldsymbol{\sigma n}$	$= \bar{\boldsymbol{t}}$	on $\partial\mathcal{B}_t$
Grad. stress	$\boldsymbol{\xi} \cdot \boldsymbol{n}$	$= \bar{\Xi}$	on $\partial\mathcal{B}_\Xi$

The classic balance of linear momentum is supplemented by an additional microforce balance. This takes into account the microstresses occurring due to the introduction of the additional gradient-contribution in the internal power density. Consequently, additional microtraction conditions for the gradient stress $\boldsymbol{\xi}$ at the GBs and at the external boundaries need to be prescribed (see also Table 3.1).

3.3 Constitutive equations with an additional grain boundary yield condition

The free energy density of the bulk, Eq. (2.22), is extended by a gradient-related defect energy $W_\mathrm{g}(\nabla\zeta)$ and, therefore, reads

$$W(\boldsymbol{\varepsilon}, \hat{\lambda}, \zeta, \nabla\zeta) = W_\mathrm{e}(\boldsymbol{\varepsilon}, \boldsymbol{\varepsilon}^\mathrm{p}(\hat{\lambda})) + W_\mathrm{h}(\zeta) + W_\chi(\zeta - \gamma_\mathrm{ac}(\hat{\lambda})) + W_\mathrm{g}(\nabla\zeta) \quad (3.8)$$

with

$$W_\mathrm{g}(\nabla\zeta) = W_0 \left(\frac{|\nabla\zeta|}{g_0}\right)^2 . \quad (3.9)$$

All remaining energy contributions are identical to Eqs. (2.24–2.26). The need for additional contributions to the free energy arises from the coarsening error made in the continuum modeling of the elastic energy. Distinct dislocation phenomena are modeled, here, with the

two contributions $W_{\mathrm{h}}(\zeta)$ and $W_{\mathrm{g}}(\nabla\zeta)$: on the one hand, the isotropic hardening resulting from SSDs, and, on the other hand, the influence of GNDs. Furthermore, $W_{\mathrm{g}}(\nabla\zeta)$ introduces a length scale into the model by means of the normalization constant g_0 (which is assumed to be a constant, here). This quadratic formulation of the defect energy results in a linear dependence of the microstress $\boldsymbol{\xi}$ on the gradient $\nabla\zeta$. The use of such a quadratic form is a convenient choice that is made in many works in the literature (e.g., Gurtin, 2000; Cermelli and Gurtin, 2002).
In Eq. (3.9), the constant W_0 is the initial defect energy. For the case of the defect energy being linear in the dislocation density, this value is often taken to be $W_0 = \bar{\alpha}G$ (cf. Hirth and Lothe, 1982; Ohno and Okumura, 2007). The shear modulus of the material is denoted by G, and $\bar{\alpha} \approx 0.5 \ldots 1$ is commonly assumed to be a constant.
In addition to the free energy density W of the bulk material, an additional energy density per unit surface is introduced on the GBs (Wulfinghoff et al., 2013)

$$W_{\Gamma}(\zeta) = \Xi_0^{\mathrm{C}}\zeta. \tag{3.10}$$

In combination with a GB yield criterion, this additional energy leads to an explicit consideration of the plastic deformation at the GBs in the continuum model. As it is shown in the following, the GB yield strength is given by Ξ_0^{C} which is assumed, here, to be constant. In a more generalized theory, this parameter would be expected to depend on the orientation of slip systems in adjacent grains as well as on the GB orientation (see, for example, the GB theory by Gurtin, 2008, which considers inter-action factors for the slip systems).
Upon neglecting thermal effects, the total dissipation, extended by the GB-contribution, reads

$$D_{\mathrm{tot}} = \mathcal{P}_{\mathrm{ext}} - \int_{\mathcal{B}} \dot{W}\,\mathrm{d}v - \int_{\Gamma} \dot{W}_{\Gamma}\,\mathrm{d}a \geq 0. \tag{3.11}$$

After exploiting $\mathcal{P}_{\text{ext}} = \mathcal{P}_{\text{int}}$, the total dissipation D_{tot} can be summarized in the form

$$D_{\text{tot}} = \int_{\mathcal{B}} \mathcal{D}\,\mathrm{d}v + \int_{\Gamma} \mathcal{D}_\Gamma\,\mathrm{d}a \geq 0. \tag{3.12}$$

In the following, the dissipation density of the bulk from Eq. (2.30) is abbreviated by $\mathcal{D} =: \mathcal{D}_\zeta$. Substitution of Eq. (3.8) in Eq. (3.11) gives a bulk dissipation, in Eq. (3.12), that reads

$$\mathcal{D} = \mathcal{D}_\zeta + \left(\boldsymbol{\xi} - \frac{\partial W_{\text{g}}}{\partial \nabla\zeta}\right) \cdot \nabla\dot{\zeta} \geq 0. \tag{3.13}$$

This inequality reduces to the dissipation inequality from Eq. (2.33) after assuming energetic stresses ($\boldsymbol{\xi} = \partial W_{\text{g}}/\partial\nabla\zeta$, consequently). If the bulk dissipation is induced by the dissipative shear stresses of the individual slip systems, see Eq. (2.37), the dissipative shear stresses are given by Eq. (2.40). Exploiting the microforce balance (i.e., $\operatorname{div}(\boldsymbol{\xi}) = \pi$) yields dissipative shear stresses of the form

$$\tau_\alpha^{\text{d}} = \tau_\alpha + \operatorname{div}(\boldsymbol{\xi}) - \beta. \tag{3.14}$$

It should be noted that an additional term, $\operatorname{div}(\boldsymbol{\xi})$, is present in Eq. (3.14), as a consequence of the consideration of the gradient $\nabla\zeta$ in the POVP. Using Eq. (3.14) in the flow rule of the bulk, Eq. (2.42), the following is obtained for the plastic slip parameters $\dot{\lambda}_\alpha$

$$\dot{\lambda}_\alpha = \dot{\gamma}_0 \left\langle \frac{\tau_\alpha^{\text{d}} - \tau_0^{\text{C}}}{\tau^{\text{D}}} \right\rangle^p = \dot{\gamma}_0 \left\langle \frac{\tau_\alpha + \operatorname{div}(\boldsymbol{\xi}) - (\tau_0^{\text{C}} + \beta)}{\tau^{\text{D}}} \right\rangle^p. \tag{3.15}$$

Furthermore, the GB dissipation from Eq. (3.11) and from Eq. (3.12), respectively, reads

$$\mathcal{D}_\Gamma = (\Xi_\Gamma - \Xi_\Gamma^{\text{e}})\,\dot{\zeta} = \Xi_\Gamma^{\text{d}}\,\dot{\zeta} \geq 0, \tag{3.16}$$

where $\Xi^{\mathrm{d}}_{\Gamma}$ is the dissipative GB microtraction. The energetic GB microtraction $\Xi^{\mathrm{e}}_{\Gamma}$ is given by

$$\Xi^{\mathrm{e}}_{\Gamma} = \partial_{\zeta} W_{\Gamma} = \Xi^{\mathrm{C}}_{0}, \tag{3.17}$$

and constitutes the GB yield strength, if GB dissipation is neglected. The GB yield function is introduced as

$$f_{\Gamma} = \Xi^{\mathrm{d}}_{\Gamma} - \Xi^{\mathrm{C}}_{\mathrm{D}} = [\![\boldsymbol{\xi}]\!] \cdot \boldsymbol{n} - (\Xi^{\mathrm{e}}_{\Gamma} + \Xi^{\mathrm{C}}_{\mathrm{D}}), \tag{3.18}$$

where $\Xi^{\mathrm{C}}_{\mathrm{D}}$ is the dissipative contribution to the GB yield strength. However, in the numerical simulations, this dissipative contribution is neglected, and, therefore,

$$f_{\Gamma} = [\![\boldsymbol{\xi}]\!] \cdot \boldsymbol{n} - \Xi^{\mathrm{e}}_{\Gamma} = [\![\boldsymbol{\xi}]\!] \cdot \boldsymbol{n} - \Xi^{\mathrm{C}}_{0}. \tag{3.19}$$

For the GBs, the Kuhn-Tucker conditions read

$$f_{\Gamma} \leq 0, \ \dot{\zeta} \geq 0, \ \dot{\zeta} f_{\Gamma} = 0, \tag{3.20}$$

assuming that the GBs behave rate-independent.

3.4 Finite element implementation

3.4.1 Linearization of the principle of virtual power

The theory at hand is implemented using the finite element method (FEM). An in-house FE-code is used for this purpose. A global Newton algorithm is used to solve for the field variables $\{\boldsymbol{u}, \zeta\}$. Thus, each FE-node has four degrees of freedom. The principle of virtual power, Eq. (3.4), is linearized for the implementation. For brevity, the subscript "n+1" of quantities of a subsequent time step is dropped.

Accordingly, the linearization of Eq. (3.4) in the virtual velocities $\delta\dot{\boldsymbol{u}}$ is given by the following form

$$\int_{\mathcal{B}} \left(\frac{\partial \boldsymbol{\sigma}}{\partial \boldsymbol{\varepsilon}}[\Delta\boldsymbol{\varepsilon}] + \frac{\partial \boldsymbol{\sigma}}{\partial \zeta}\Delta\zeta \right) \cdot \delta\dot{\boldsymbol{\varepsilon}}\,\mathrm{d}v = -\int_{\mathcal{B}} \boldsymbol{\sigma} \cdot \delta\dot{\boldsymbol{\varepsilon}}\,\mathrm{d}v + \int_{\partial\mathcal{B}_{\mathrm{t}}} \bar{\boldsymbol{t}} \cdot \delta\dot{\boldsymbol{u}}\,\mathrm{d}a \quad \forall\delta\dot{\boldsymbol{u}}, \tag{3.21}$$

and the linearization of Eq. (3.4) in $\delta\dot{\zeta}$ is

$$\begin{aligned} &\int_{\mathcal{B}} \left(\delta\dot{\zeta}\Big(\frac{\partial \beta}{\partial \zeta} - \frac{\partial \check{p}}{\partial \zeta}\Big)\Delta\zeta - \delta\dot{\zeta}\,\frac{\partial \check{p}}{\partial \boldsymbol{\varepsilon}} \cdot \Delta\boldsymbol{\varepsilon} + \frac{\partial \boldsymbol{\xi}}{\partial \nabla\zeta}[\nabla(\Delta\zeta)] \cdot \nabla(\delta\dot{\zeta}) \right) \mathrm{d}v \\ &\qquad + \int_{\Gamma_{\mathrm{act}}} \delta\dot{\zeta}\,\frac{\partial \Xi^{\mathrm{e}}_{\Gamma}}{\partial \zeta}\Delta\zeta\,\mathrm{d}a = -\int_{\mathcal{B}} \left(\Big(\beta - \check{p}\Big)\delta\dot{\zeta} + \boldsymbol{\xi} \cdot \nabla(\delta\dot{\zeta}) \right) \mathrm{d}v \\ &+ \int_{\partial\mathcal{B}_{\Xi}} \bar{\Xi}\,\delta\dot{\zeta}\,\mathrm{d}a - \underbrace{\int_{\Gamma\backslash\Gamma_{\mathrm{act}}} (\Xi^{\mathrm{d}}_{\Gamma} + \Xi^{\mathrm{e}}_{\Gamma})\,\delta\dot{\zeta}\,\mathrm{d}a}_{I} - \underbrace{\int_{\Gamma_{\mathrm{act}}} (\Xi^{\mathrm{C}}_{\mathrm{D}} + \Xi^{\mathrm{e}}_{\Gamma})\,\delta\dot{\zeta}\,\mathrm{d}a}_{J}\,. \end{aligned} \tag{3.22}$$

Here, the algorithmic tangent is denoted by the partial derivatives. In Eq. (3.22), it has been exploited, already, that the yield condition is fulfilled on the set of plastically active GB-nodes, i.e., on $\Gamma_{\mathrm{act}} = \{\boldsymbol{x} \in \Gamma : \dot{\zeta} > 0\}$. It is remarked that the integral I is not relevant for the computations if $\delta\dot{\zeta}$ is set to vanish on $\Gamma\backslash\Gamma_{\mathrm{act}}$. For the same reason, the linearization of I can be neglected.
The linearized form, Eqs. (3.21, 3.22), is discretized in space with the FEM. An integration point routine is used to compute the stresses and the tangent during each Newton iteration. Therefore, an implicit Euler-scheme is applied to Eq. (3.15) which yields

$$\lambda_\alpha = \lambda_{\alpha,\mathrm{n}} + \dot{\gamma}_0\Delta t \left\langle \frac{\boldsymbol{\sigma} \cdot \boldsymbol{M}^{\mathrm{S}}_{\alpha} - \check{p} - \tau^{\mathrm{C}}_{0}}{\tau^{\mathrm{D}}} \right\rangle^{p}. \tag{3.23}$$

For the local Newton scheme, the residual is given by

$$\boldsymbol{r}^{\sigma} = -\mathbb{C}^{-1}[\boldsymbol{\sigma}] + \boldsymbol{\varepsilon} - \boldsymbol{\varepsilon}^{\mathrm{p}}_{\mathrm{n}} - \Delta\boldsymbol{\varepsilon}^{\mathrm{p}} = \boldsymbol{0} \tag{3.24}$$

with the increment $\Delta\boldsymbol{\varepsilon}^{\mathrm{p}} = \sum_\alpha \Delta t \dot{\gamma}_0 \left\langle (\tau_\alpha - \check{p} - \tau_0^{\mathrm{C}})/\tau^{\mathrm{D}} \right\rangle^p \boldsymbol{M}_\alpha^{\mathrm{S}}$, and by the additional contribution

$$r^{\mathrm{p}} = \gamma_{\mathrm{ac,n}} + \Delta\gamma_{\mathrm{ac}} - \zeta - \frac{\check{p}}{H_\chi} = 0 \tag{3.25}$$

with $\Delta\gamma_{\mathrm{ac}} = \sum_\alpha \Delta t \dot{\gamma}_0 \left\langle (\tau_\alpha - \check{p} - \tau_0^{\mathrm{C}})/\tau^{\mathrm{D}} \right\rangle^p$. Both Eqs. (3.24, 3.25) are solved for the stresses $\boldsymbol{\sigma}$ and $\check{p}$ using the enhanced Newton algorithm by Wulfinghoff and Böhlke (2013). For the global problem, the algorithmic tangent needs to be computed (see Eqs. (3.21, 3.22)) at each integration point during each (global) iteration. In order to determine the algorithmic tangent, the total derivatives of the residuals are set to vanish, in the following. In detail, this gives

$$\mathrm{d}\boldsymbol{r}^\sigma = \frac{\partial \boldsymbol{r}^\sigma}{\partial \boldsymbol{\sigma}}[\mathrm{d}\boldsymbol{\sigma}] + \mathbb{I}^{\mathrm{s}}[\mathrm{d}\boldsymbol{\varepsilon}] + \frac{\partial \boldsymbol{r}^\sigma}{\partial \check{p}}\,\mathrm{d}\check{p} = \boldsymbol{0}, \tag{3.26}$$

$$\mathrm{d}r^{\mathrm{p}} = \frac{\partial r^{\mathrm{p}}}{\partial \boldsymbol{\sigma}} \cdot \mathrm{d}\boldsymbol{\sigma} + \frac{\partial r^{\mathrm{p}}}{\partial \check{p}}\,\mathrm{d}\check{p} + \frac{\partial r^{\mathrm{p}}}{\partial \zeta}\,\mathrm{d}\zeta = 0. \tag{3.27}$$

Accordingly, in matrix-vector notation, this reads

$$\underbrace{\begin{pmatrix} -\dfrac{\partial \hat{r}^\sigma}{\partial \hat{\sigma}} & -\dfrac{\partial \hat{r}^\sigma}{\partial \check{p}} \\ -\dfrac{\partial r^{\mathrm{p}}}{\partial \hat{\sigma}} & -\dfrac{\partial r^{\mathrm{p}}}{\partial \check{p}} \end{pmatrix}}_{\hat{A}} \begin{pmatrix} \mathrm{d}\hat{\sigma} \\ \mathrm{d}\check{p} \end{pmatrix} = \underbrace{\begin{pmatrix} \hat{I} & 0 \\ 0 & -1 \end{pmatrix}}_{\hat{B}} \begin{pmatrix} \mathrm{d}\hat{\varepsilon} \\ \mathrm{d}\zeta \end{pmatrix}, \tag{3.28}$$

and, thus,

$$\begin{pmatrix} \mathrm{d}\hat{\sigma} \\ \mathrm{d}\check{p} \end{pmatrix} = \underbrace{\hat{A}^{-1}\hat{B}}_{\hat{D}} \begin{pmatrix} \mathrm{d}\hat{\varepsilon} \\ \mathrm{d}\zeta \end{pmatrix}, \tag{3.29}$$

$$\hat{D} = \begin{pmatrix} \dfrac{\partial \hat{\sigma}}{\partial \hat{\varepsilon}} & \dfrac{\partial \hat{\sigma}}{\partial \zeta} \\ \dfrac{\partial \check{p}}{\partial \hat{\varepsilon}} & \dfrac{\partial \check{p}}{\partial \zeta} \end{pmatrix}. \tag{3.30}$$

It is noted that the matrix $\hat{D} \in Sym$ due to the symmetry of the components $\partial \boldsymbol{r}^\sigma / \partial \boldsymbol{\sigma} \in Sym$ and $\partial \boldsymbol{r}^\sigma / \partial \check{p} = \partial r^{\mathrm{p}} / \partial \boldsymbol{\sigma}$.

3.4.2 Grain boundary discretization

The GB integral J over Γ_{act} in Eq. (3.22) is approximated by

$$- \int_{\Gamma_{\mathrm{act}}} (\Xi_{\mathrm{D}}^{\mathrm{C}} + \Xi_{\Gamma}^{\mathrm{e}}) \delta \dot{\zeta} \, \mathrm{d}a \cong - \sum_{i \in \Gamma_{\mathrm{act}}} (\Xi_{\mathrm{D}}^{\mathrm{C}} + \Xi_{\Gamma}^{\mathrm{e}})_i w_i \delta \dot{\zeta}_i, \tag{3.31}$$

where the integration points are identified by i and the quadrature weights by w_i. The integration points are chosen to coincide with the nodes of the FE mesh. This approach is motivated by the work of Liebe and Steinmann (2001) and eases the determination of the plastically active part Γ_{act} of the GBs. It is remarked that this is a well-established strategy for linear FE-elements which are used throughout the present work. The discretized counterpart of Γ_{act} is given by the active set $\mathcal{A}$ of GB nodes where $f_\Gamma = 0$ and $\Delta \zeta \geq 0$. Plastically inactive GB nodes are treated as microhard nodes using a Dirichlet condition, i.e., $\Delta \zeta = 0$ and $\delta \dot{\zeta} = 0$. Therefore, the computation of the integral

$$- \int_{\Gamma \backslash \Gamma_{\mathrm{act}}} (\Xi_{\mathrm{d}} + \Xi_{\Gamma}^{\mathrm{e}}) \, \delta \dot{\zeta} \, \mathrm{d}a \cong - \sum_{i \in \Gamma \backslash \Gamma_{\mathrm{act}}} (\Xi_{\mathrm{d}} + \Xi_{\Gamma}^{\mathrm{e}})_i w_i \delta \dot{\zeta}_i, \tag{3.32}$$

in Eq. (3.22), is neither necessary nor possible. The associated rows and columns are eliminated both from the global stiffness matrix and from the residual. Since Eq. (3.32) is neglected during the assembly, the corresponding components of the force residual do not vanish when convergence is reached but take the values $(\Xi_{\mathrm{d}} + \Xi_{\Gamma}^{\mathrm{e}})_i w_i$, instead. This result is exploited in order to evaluate the yield criterion at the inactive nodes and update the active set $\mathcal{A}$. The algorithm to determine $\mathcal{A}$, and the contribution of the GB term in Eq. (3.31) to the global stiffness matrix as well as to the force residual is summarized in Table 3.2.

Table 3.2: Algorithm for determination of plastic activity of grain boundary nodes. Table reprinted from Wulfinghoff et al. (2013) with permission from Elsevier.

1. Grain boundary contribution to the residual and the stiffness matrix:
   For all GB nodes in active set $\mathcal{A}$:
   a) Add $-(\Xi^{\mathrm{e}}_{\Gamma} + \Xi^{\mathrm{C}}_{\mathrm{D}})_i w_i$ to the right-hand side of the global force residual (cf. Equations (3.22) and (3.31))
   b) Add $(\partial(\partial W_{\Gamma}/\partial\zeta)/\partial\zeta)_i w_i$ to the diagonal of the global stiffness matrix (If W_{Γ} is assumed to be linear in ζ, the stiffness matrix, consequently, does not need to be updated)
2. Update of active set $\mathcal{A}$:
   If `res`*<`tol`$_{\mathcal{A}}$ (i.e., update $\mathcal{A}$ only if close to convergence)
   a) For all GB nodes i:
      If $i \in \mathcal{A}$
         If $\Delta\zeta_i = \zeta_i - \zeta_{i,\mathrm{n}} < 0$
         - $\mathcal{A} \leftarrow \mathcal{A} \backslash \{i\}$
         - $\zeta_i = \zeta_{i,\mathrm{n}}$

      Else if $i \notin \mathcal{A}$
         If $f_{\Gamma} = \Xi_{\mathrm{d}} - \Xi^{\mathrm{C}}_{\mathrm{D}} > 0$
         - $\mathcal{A} \leftarrow \mathcal{A} \cup \{i\}$

   b) If active set $\mathcal{A}$ has changed:
      Recompute the stiffness matrix and the residuals (including step "1.")
3. If `res`*<`tol`$_{\mathrm{force}}$ and $\mathcal{A}$ has not changed the time step is considered converged

*: `res` is the maximum norm of the force residual

3.5 Gradient plasticity simulations

3.5.1 Simulation setup, boundary conditions, and discretization details

Tensile test simulations are carried out using grain aggregates of simplified shape. For the average grain sizes, values are chosen based on the experimental data of copper microwires by Yang et al. (2012). The crystal orientations of the grains are assigned randomly, approximately uniform due to lack of availability of orientation data in the experimental reference. In Table 3.3, the microstructural features of the microwires are shown. The grain aggregate dimensions and the chosen average grain sizes for the simulations are listed, additionally. In Fig. 3.1a, the corresponding discretizations of the simulation volumes are depicted and the boundary conditions for the simulation setup are shown in Fig. 3.1b. Lateral contraction is allowed for and the displacements in the loading direction are prescribed. Initially, the simulation volume is undeformed, i.e., all values of $\gamma_{\rm ac}$ and ζ are set to zero.

Table 3.3: Microstructural features of copper microwires, coarse-grained at different annealing temperatures (AT) ([1]: data from Yang et al., 2012), and dimensions of simplified grain aggregates used in FE simulations. Average grain sizes (GS) determined by half of the summed mean values of the longitudinal and transversal average grain sizes.

Experiments[1]			Simulations	
Wire diameter[1] $D_{\rm exp}$ in µm	Avg. GS[1] $d_{\rm avg,exp}$ in µm	AT[1] in °C	Width L in µm	Avg. GS in µm
25	5.2 ± 1.0	400	25	6.25
25	11.5 ± 2.3	600	25	12.5
25	22 ± 3	900	25	18.75

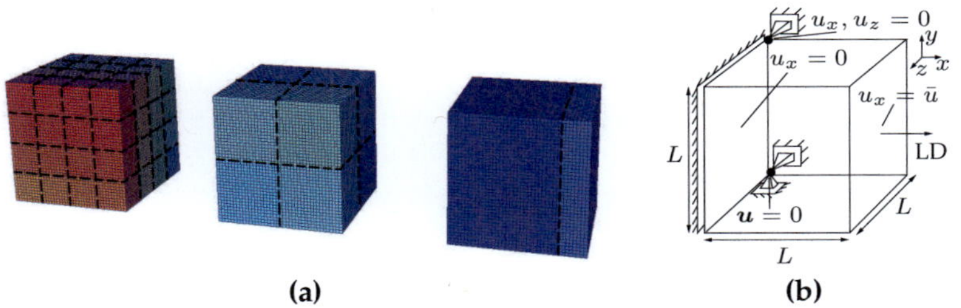

Figure 3.1: (a) Discretizations of simplified grain aggregates used for simulating the mechanical response of annealed copper microwire tensile tests from Yang et al. (2012). From left to right: Simplified grain aggregates used for microwires with different grain sizes. Grain boundaries are highlighted by dashed lines. (b) Boundary conditions for tensile test simulations. LD: loading direction. Fig. (a) reprinted from Wulfinghoff et al. (2013) with permission from Elsevier, and Fig. (b) similar to Wulfinghoff et al. (2013).

The boundary $\partial\mathcal{B}$ is treated as a Neumann boundary using $\bar{\Xi} = 0$. For all discretizations, standard trilinear hexahedrons are utilized. The number of finite elements used is 64000 (cf. Fig. 3.1) giving a total of 275684 DOFs for the simulations. This discretization is chosen based on a convergence study for all three grain aggregates as a compromise between accuracy and manageable computational times (see Fig. F.1.1 for details). Adaptive time step-control is used during the simulation time of 1 s. The global Newton scheme is considered to be converged if the set of active GB nodes $\mathcal{A}$ has not been changed during an iteration and if the initial residual has been reduced at least by a factor of 10^{-8} (i.e., $\mathrm{tol}_{\mathrm{force}} = 10^{-8}\,\mathrm{res}_0$, see Table 3.2). For the active set search, the chosen tolerance value is $\mathrm{tol}_{\mathcal{A}} = 10\,\mathrm{tol}_{\mathrm{force}}$ (see also Table 3.2), and for the integration point subroutine the tolerance 10^{-8} is used (associated to the maximum norm of the local residual, Eqs. (3.24, 3.25)). Residual values of two characteristic time steps are shown in Table 3.4. The active-set search on the GBs leads to additional iterations which mainly occur in the elastic-plastic transition regime of the simulations (see the highlighted residuals in Table 3.4).

Table 3.4: Residuals for two typical time steps: one with active set search (in the elastic-plastic transition regime), and one without active set search (in the plastic regime) for 64 grains. Table reprinted from Wulfinghoff et al. (2013) with permission from Elsevier.

Δt / res_0	Maximum norm of residual								
$0.01s$	Iteration 1	2	3	4	5	6	7	8	9
	5.67	6.70	7.96e-01	9.90e-02	1.75e-02	4.62e-04	**5.93e-07**	6.69	1.15
1.84e+03	10	11	**12**	13	14	**15**	16	17	
	1.20e-01	3.09e-03	**5.11e-06**	1.34	1.25e-02	**2.99e-06**	1.14e-03	2.24e-09	
$0.32s$	Iteration 1	2	3	4	5	6	7		
5.88e+04	1.52e+01	9.92	2.83	4.92e-01	3.60e-02	1.21e-03	5.29e-06		

Once all GB nodes are plastically active, however, additional iterations do not occur further. An overview on the computational times of the simulations in this thesis, with and without GB contributions, is given in Table D.1.

3.5.2 Influence of the model parameters on the mechanical response

The influence of the model parameters is discussed in this section, considering the basic parameters that have already been presented in Table 2.2. For the anisotropic elastic constants of copper, values are taken from Rösler et al. (2006). At first, size-independent (classic) plasticity is considered, i.e., the GB and gradient influences on the mechanical response are neglected. In Fig. 3.2, it is shown that, then, a decrease in the edge length L of the cubic oligocrystal does not alter the mechanical response and the model behaves size-independent. It is noted that, for sufficiently large values of the strain rate sensitivity p, the material model approximates rate-independent behavior. A moderate variation of the reference strain rate $\dot{\gamma}_0$ has no noticeable effect, in this case. In the rate-independent limit, the offset stress (i.e., the yield stress of the lower three (coinciding) curves in Fig. 3.2) is mainly controlled by $\tau_0^{\mathrm{C}} + \tau^{\mathrm{D}}$. The hardening parameters Θ and τ_∞^{C} of the used Voce-approach can be used to control the evolution of the hardening slope. If, however, the GB yield strength Ξ_0^{C} and the ratio of W_0/g_0^2 are non-zero, the model behaves size-dependent. A decrease in L leads to a stiffer overall mechanical response see (Fig. 3.2).

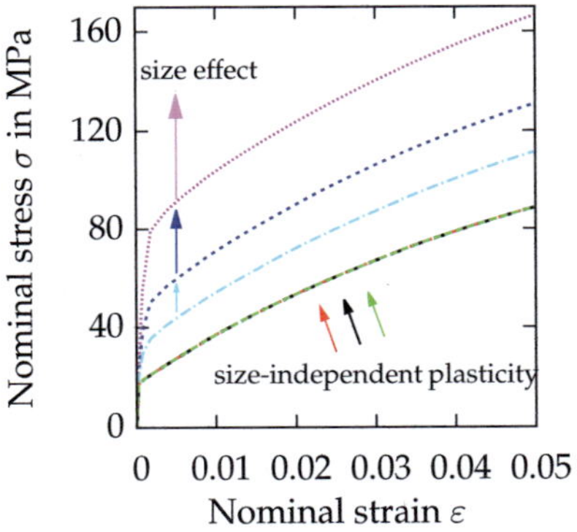

	Ξ_0^{C} (N/m)	g_0 (1/μm)	L (μm)
········	50	1.92	6.25
····	50	1.92	12.5
-·-·-	50	1.92	25
-·-···	0	$W_0/g_0^2 = 0$	6.25
·· ··	0	$W_0/g_0^2 = 0$	12.5
----	0	$W_0/g_0^2 = 0$	25

Figure 3.2: Influence of the structure size on the mechanical tensile response of an 8-grain aggregate in the presence and absence of grain boundary and gradient influences. Basic figure reprinted from Wulfinghoff et al. (2013) with permission from Elsevier.

It is noted that the model formulation in Wulfinghoff et al. (2013) is slightly different, using only one parameter K_G instead of the two parameters W_0 and g_0 in the present work. Therefore, the values of the normalization constant g_0 shown in this section are obtained by converting the values of K_G through the correspondence of

$$g_0 = \sqrt{2W_0/K_G} \tag{3.33}$$

with $W_0 = 0.5G$ and $G = E/(2 + 2\nu)$ with an exemplary isotropic Young's modulus of $E = 100$ GPa for copper. The influence of the GB yield strength Ξ_0^{C} and the normalization constant g_0 are contrasted in Fig. 3.3. Varying g_0 (and, thus, the internal length scale) over three orders of magnitude, for a constant value of Ξ_0^{C}, only slightly changes the overall mechanical response.However, the elastic plastic-transition regime is affected, as smaller values of g_0 lead to a more abrupt transition behavior, compared to larger values (see zoomed-in image in Fig. 3.3). It should be noted that the simulation results for $g_0 = 19.2/\mu\mathrm{m}$ in this section are only regarded as a trend, due to a need for more refined meshes for this value of g_0 in combination with the value of W_0. Varying the GB yield strength over three orders of magnitude, however, leads to a pronounced increase in the overall yield strength (Fig. 3.3). Thus, mainly the GB yield strength Ξ_0^{C} controls the overall yield stress. Consequently, Ξ_0^{C} needs to be fitted to stress-strain curves obtained in the elastic-plastic regime for specimen with different diameters and constant grain size or constant diameter and different grain sizes, respectively. Possible data sources are experimental investigations as well as discrete dislocation dynamics simulations.

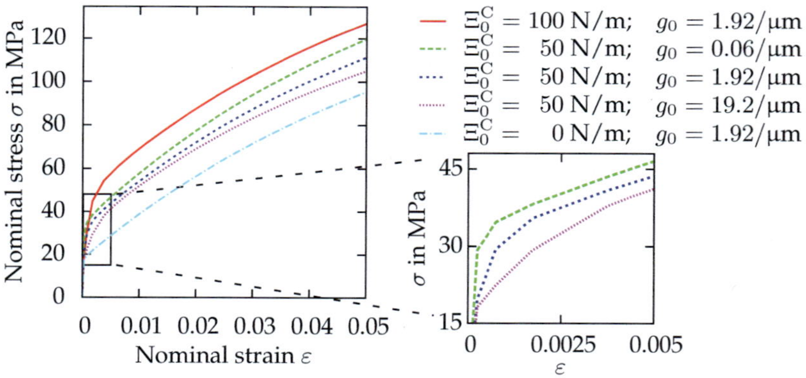

Figure 3.3: Influence of Ξ_0^{C} and g_0, respectively, on the mechanical tensile response of an 8-grain aggregate. Basic figure reprinted from Wulfinghoff et al. (2013) with permission from Elsevier. Zoom image: influence of g_0 on the elastic-plastic transition regime.

3.5.3 Comparison of simulation results to experimental data

One of the aims of the present thesis is the prediction of size effects on microspecimen. Therefore, an experimental study on copper microwires by Yang et al. (2012) is considered as a reference to test the GP model. The model parameters $\tau_0^C, \Theta, \tau_\infty^C$ as well as the GB and gradient-related parameters Ξ_0^C, g_0 are fitted by comparison of simulation results to the experimental stress-strain data. Thereby, the set of parameters, which has already been presented in Table 2.2, is obtained. The GB yield strength is $\Xi_0^C = 55$ N/m and the value used for the normalization constant is $g_0 = 1.92/\mu$m. Both the experimental reference and the model response are contrasted in Fig. 3.4. Smaller deviations from the hardening behavior of the microwires occur for different grain sizes in the model response. Overall, however, the experimental reference is reproduced comparably well. Notably, the overall size-effect that is observed in the experimental data is accounted for by the simulations.

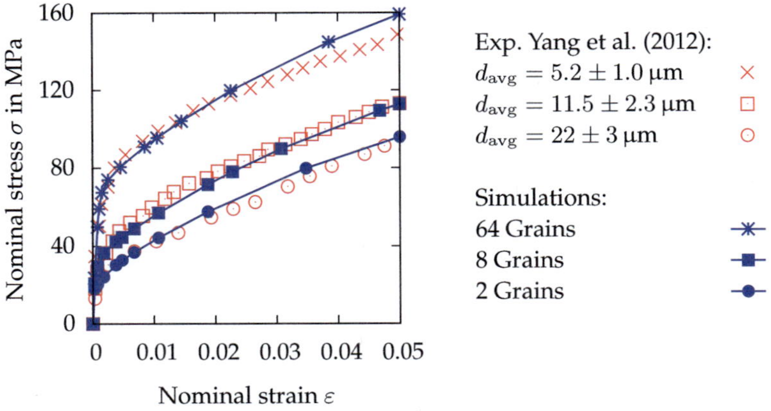

Figure 3.4: Comparison of simulation results of simplified grain aggregates to experimental data of copper microwire tensile tests by Yang et al. (2012). The average grain size of the specimens is indicated by d_{avg}. Simulation data from Wulfinghoff et al. (2013).

In addition, the distributions of the micromorphic approximation of the accumulated plastic slip at an engineering strain of $\varepsilon = 0.05$ are depicted for all three simulated grain-aggregates in Fig. 3.5. It is recalled that the use of the micromorphic approximation of γ_{ac} in the simulations is feasible if the penalty parameter H_χ is chosen large enough and a sufficiently fine FE-mesh is used. The used value of the penalty parameter throughout this work is $H_\chi = 10^8$ MPa. A comparison of both field distributions for the oligocrystalline case of this section is shown in Appendix C.2.

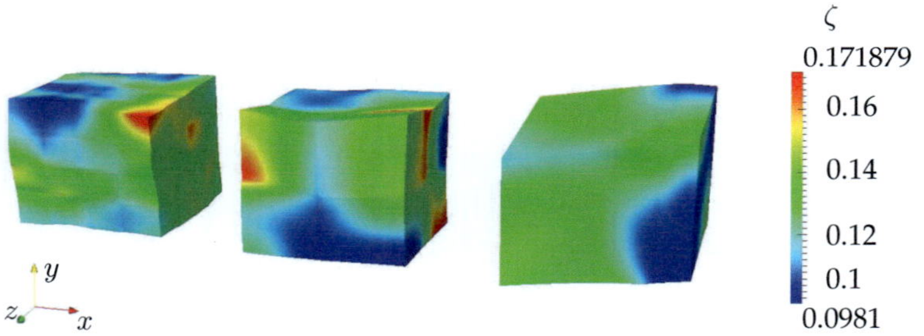

Figure 3.5: Resulting distributions of the micromorphic variable ζ for tensile test simulations of simplified grain aggregates, displacement scaled by a factor of three. Figure reprinted from Wulfinghoff et al. (2013) with permission from Elsevier.

At large parts of the GBs, in Fig. 3.5, the micromorphic variable shows smaller values than in the adjacent grains (see also Fig. 3.1a for the location of the GBs). The GB yield strength acts as additional resistance to plastic flow, there, compared to the grain interior. This mechanism is shown in more detail in Fig. 3.6 for the 8-grain-aggregate. During the early stages of plastic deformation, the bulk nodes become plastically active but at the GB nodes plasticity is still prevented (see time step t_1 in Fig. 3.6). Once the GB yield condition is fulfilled at these nodes, the GB nodes also become plastically active (see time steps t_2–t_4 in Fig. 3.6). Additionally, the evolution of the gradient hardening stress $-\mathrm{div}\,(\boldsymbol{\xi})$ is shown in Fig. 3.6. This quantity increases close to the GBs during the evolution of the plastic deformation. In the beginning of the plastic deformation, the rate of increase of $-\mathrm{div}\,(\boldsymbol{\xi})$ is higher than at the end (see time steps t_1–t_2 in Fig. 3.6). Once the GBs start yielding, however, the increase of $-\mathrm{div}\,(\boldsymbol{\xi})$ saturates (see time steps t_3–t_4 in Fig. 3.6).

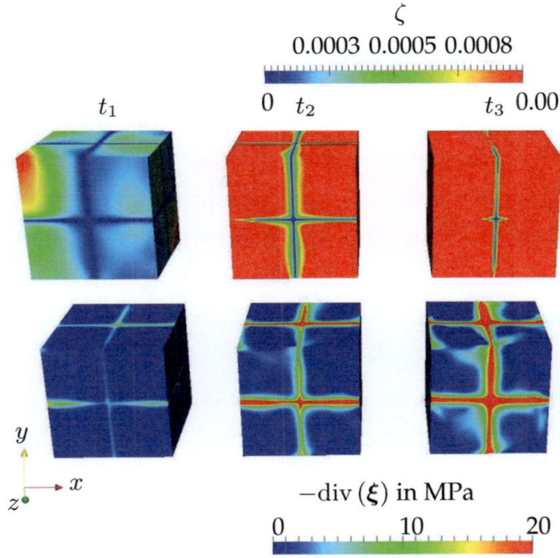

Figure 3.6: Distributions of the micromorphic variable ζ and the gradient hardening stress $-\mathrm{div}\,(\boldsymbol{\xi})$ for tensile test simulations of a simplified grain aggregate with 8 grains. Depicted are the distributions at the end of each of the first four time steps. The color scale of the upper image sequence is limited in order to visualize the grain boundary yielding. Basic figure from Wulfinghoff et al. (2013), reprinted with permission from Elsevier.

3.6 Gradient plasticity simulations considering the crystal orientations of the grains in a simplified manner

3.6.1 Motivation

In the recent experimental work by Chen et al. (2015), the deformation behavior of oligocrystalline gold microwires with varying diameters was investigated for both uniaxial tension and torsion loading. Contrary size effects were observed for the two different loading cases. In the microstructural studies of this work it was found that the microwires of different thickness exhibit also different average grain sizes and textures, respectively. Consequently, it was assumed that both microstructural characteristics influence the resulting size effects of the overall mechanical response. However, a confirmation of these assumed influences was only established for the influence of the average grain size. In the present section of this thesis, the size-dependent mechanical response of the microwires is modeled with the GP theory from Sections 3.2–3.4. By FE simulations of oligocrystalline grain aggregates, the influence of the texture on the size effects is investigated under both loading conditions in a simplified manner. It is shown that the, experimentally observed, contrary size effects can only be reproduced if the individual textures of the microwires of different thickness are considered in the modeling. The content of this section is largely taken from Bayerschen et al. (2016b).

3.6.2 Summary of the experimental characterization of the microwires

The experimental part of this work was performed by M. Ziemann, M. Walter and P.A. Gruber (Institute of Applied Materials, KIT). In the

following, a brief summary of the experimental evaluation is given.
A quantitative examination of the grain orientation distribution in the cross-sections of the microwires was performed for four selected microwire diameters in the range of 15 µm – 60 µm. Therefore, up to four cross-sections of the respective microwires from Chen et al. (2015) were reanalyzed. The original electron backscatter diffraction (EBSD) maps from the investigated wires (see, for example, Fig. 3.7a and Figs. 3.7d–3.7f) were evaluated with MATLAB R2014a and MTEX toolbox 4.1.1 (see Hielscher and Schaeben, 2008; Bachmann et al., 2010).

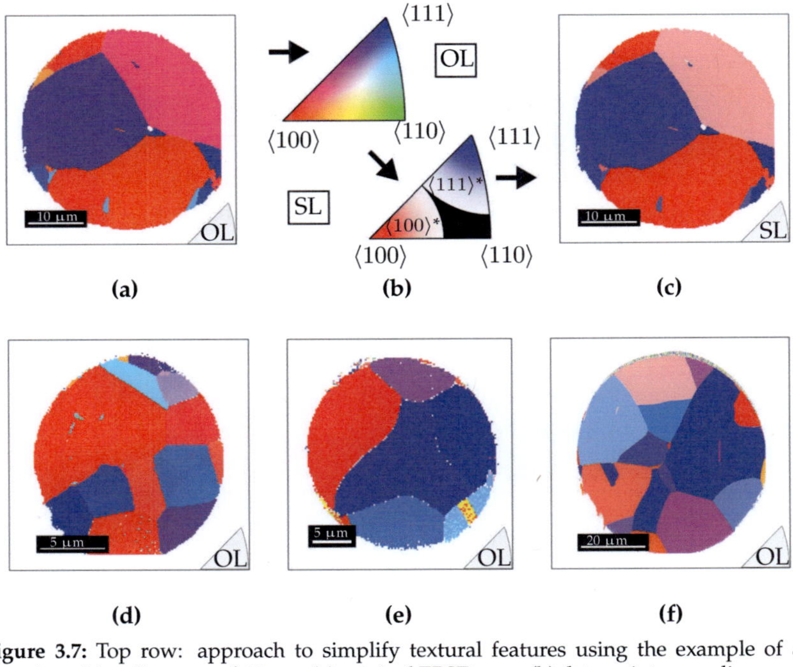

Figure 3.7: Top row: approach to simplify textural features using the example of an Au-wire with a diameter of 40 µm: (a) original EBSD map, (b) data points revealing a misorientation of $\leq 27.37°$ related to either the ⟨100⟩-axis or to the ⟨111⟩-axis, respectively, are defined as ⟨100⟩*- or ⟨111⟩*-oriented spots (data points located outside the defined areas are ignored), (c) simplified microstructure. Bottom row: original EBSD maps from wires with a diameter of (d) 15 µm, (e) 25 µm, and (f) 60 µm. (OL: original legend, SL: simplified legend). Figures reprinted from Bayerschen et al. (2016b) with permission from Springer. Images courtesy of M. Ziemann.

For a simplification of the GP simulation approach, the wires are assumed to only consist of ideally $\langle 100 \rangle$- and $\langle 111 \rangle$-oriented grains. Therefore, the experimental data were prepared such that the orientation of each spot within a cross section was counted either as a $\langle 100 \rangle^*$- or as a $\langle 111 \rangle^*$-oriented normal to the cross-section surface when its misorientation is located within a radius of $27.37°$, related to either one of the both main directions (see Figs. 3.7b–3.7c). The cut-off value of $27.37°$ was chosen based on the fact that it is half the angle between the $\langle 100 \rangle$- and the $\langle 111 \rangle$-axis. Data points with orientations outside of the two resulting circles are rare for the investigated samples and, thus, negligible for the present work. Then, by counting the number of spots for both orientations, the area shares of the $\langle 100 \rangle^*$- as well as the $\langle 111 \rangle^*$-oriented areas were determined. Finally, the average area shares were calculated for every microwire diameter. These area shares are listed in Table 3.5, in addition to the average grain sizes (determined by using a modified line-interception method and already published in Chen et al., 2015).

Table 3.5: Microstructural features of coarse-grained Au microwires with different diameter ([1]: values from Chen et al., 2015). Data and table reprinted from Bayerschen et al. (2016b) with permission from Springer.

Nom. wire diameter D (µm)	Real wire diameter[1] D_{exp} (µm)	Avg. grain size[1] $d_{\text{avg,exp}}$ (µm)	Area share $\langle 111 \rangle^*$ (%)	Area share $\langle 100 \rangle^*$ (%)
15	14.78	2.96	33	67
25	24.46	3.49	45	55
40	37.38	8.63	67	33
60	59.53	6.38	75	25

Considering the individual area shares of the simplified microstructures from Table 3.5 in comparison to the color maps of the related original EBSD maps, one can suppose that the calculated values are matching comparably well for the two wires of smaller diameter (e.g., Fig. 3.7d). In contrast, the simplified grain orientation distributions of the two wires with bigger diameter do not seem to represent the given orientations precisely enough (e.g., Fig. 3.7f). Instead, larger fractions of grains with stronger misorientations related to the ideal $\langle 100\rangle$-axis or the ideal $\langle 111\rangle$-axis are present. Both findings have also been confirmed for the other cross-sections of the corresponding microwires (see Bayerschen et al., 2016b). The microwires of smaller diameters, however, mainly consist of almost ideally $\langle 100\rangle$- and $\langle 111\rangle$-oriented grains (comparing, for example, Fig. 3.7a to Fig. 3.7f). Therefore, additional gradient plasticity simulations are carried out for the two thicker microstructures using equal fractions of $\langle 100\rangle$- and $\langle 111\rangle$-oriented grains. Thereby, the influence of the not adequately enough described microstructure on the size effect can be investigated.

3.6.3 Finite element simulation setups and discretizations

Finite element simulations are performed and the overall mechanical responses are compared to the experimentally determined mechanical responses of the microwires in Chen et al. (2015). The used BCs for the tensile loading are identical to the BCs described in Section 3.5.1, and the used BCs for the torsion loading are identical to the BCs described in Section 2.5.3. For the tensile loading, a displacement is imposed on the top surface, until a nominal strain of $\varepsilon = 0.05$ is reached. It is recalled that lateral contraction is allowed for by the BCs in the transversal directions. For the torsion loading, the top and bottom surfaces (located in direction of the central specimen axis) are twisted in opposing directions until a maximum shear of $\gamma_{r=R} = 0.02$ is reached. On this account, it is noted that the displacement in the direction of the central specimen axis is free, except at one FE node.
In the beginning of the simulation, the accumulated plastic slip and the micromorphic variable are set to vanish in the whole simulation volume.

The GB yield strength is then assigned to the respective nodes of the element surfaces coinciding with the GB planes. On the surface of the simulation volume, microfree conditions are imposed for the accumulated plastic slip, i.e., the accumulated plastic slip is not restricted by the BCs, there. In the beginning of the deformation, all nodes on the GBs are set to behave microhard. Thus, the GBs are plastically inactive until the GB yield condition is fulfilled and the GB nodes become plastically active. The volume length L_0 for each diameter is chosen such that $L_0/D = 1.25 = const.$, in order to allow for a comparison of the results obtained for different diameters at the same values of the overall maximum shear (see Table 3.6 for the diameters D).

Table 3.6: Discretizations of oligocrystalline gold microwires. Abbreviations used are: DOFs - degrees of freedom, ECD - equivalent circle diameter, FEs - finite elements. Table reprinted from Bayerschen et al. (2016b) with permission from Springer.

Diameter D_{sim} (µm)	Average grain size ECD: $d_{\mathrm{avg,sim}}$ (µm)	FEs	DOFs
15	3.67	54720	235396
25	6.12	54864	236180
40	9.80	54720	235396
60	14.69	54864	236180

Exemplary, the grain aggregate with smallest diameter is investigated in a convergence study under tensile as well as under torsion loading. Based on these results, the discretizations of the simplified grain aggregates are chosen such that an approximate doubling of the DOFs of the FE mesh leads to a reduction of less than 1.5% in the relative error of the overall mechanical responses after the final time step of the grain aggregate simulations. The computational times remain within the order of several hours for the employed discretizations, listed in

Table 3.6. Due to the FE-discretization algorithm in the used mesh-generator (in ABAQUS), slightly different FE numbers are present in the individual discretizations. However, no substantial influence on the overall mechanical response is observable by the resulting small differences in the DOFs of the grain aggregates.

3.6.4 Modeling of the microstructure

For modeling the microstructure of the wires, grains of simplified shape are considered (see Fig. 3.8a). The GBs are discretized by partitioning the simulation volume such that the GBs are composed of the planar surfaces of the elements in the FE mesh. Therefore, equidistant positioning of three GB planes is performed in each spatial direction.

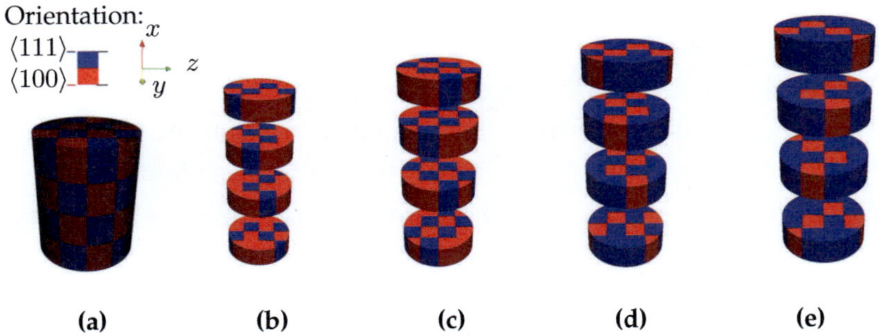

Figure 3.8: Textures of oligocrystalline grain aggregates used in finite element simulations: (a) equally distributed (alternating) ⟨100⟩- and ⟨111⟩-orientations, depicted are the Gauss-point volumes of the employed finite element mesh, (b) 15 µm-aggregate with 20 ⟨111⟩-grains and 44 ⟨100⟩-grains, (c) 25 µm-aggregate with 24 ⟨111⟩-grains and 40 ⟨100⟩-grains, (d) 40 µm-aggregate with 44 ⟨111⟩-grains and 20 ⟨100⟩-grains, and (e) 60 µm-aggregate with 48 ⟨111⟩-grains and 16 ⟨100⟩-grains. The size of all grain-aggregate images has been rescaled for the visualization. The specific number of ⟨100⟩- and ⟨111⟩-oriented grains for (b) to (e) are adapted to the experimental data in Table 3.5. Figure reprinted from Bayerschen et al. (2016b) with permission from Springer.

Motivated by the observed average grain sizes of the oligocrystalline microwire specimens (see Table 3.5), 16 grains are modeled within each of four grain layers per aggregate, i.e., the whole simulation volume is composed of 64 grains. The resulting average grain sizes in terms of the equivalent circle diameter (ECD) are listed in Table 3.6.
Each grain is assigned either a $\langle 100\rangle$- or a $\langle 111\rangle$-orientation with regard to the specimen coordinate system (see also Fig. 3.8). It is recalled that this means that, for $\langle 100\rangle$-grains, the $\langle 100\rangle$-crystal directions are aligned with the Cartesian coordinate system while, for $\langle 111\rangle$-grains, the $\langle 111\rangle$-crystal direction is aligned with the central specimen axis. At first, it is assumed that equal area shares of both orientations are present in the cross-sections and the orientations are assigned to the grains such that $\langle 100\rangle$- and $\langle 111\rangle$-orientations are alternating (see Fig. 3.8a).
However, based on the reevaluation of the orientation distribution of several cross-sections from the experiments, a trend towards more $\langle 100\rangle$-grains in the investigated cross-sections is observable for smaller wire diameters. Therefore, the texture of the microwires is accounted for in the simulations, subsequently. For each wire diameter, the crystal orientations of several grains within each cross section are switched, for example, from the $\langle 111\rangle$- to the $\langle 100\rangle$-orientation, until the area shares of the $\langle 100\rangle$- and the $\langle 111\rangle$-orientation are approximately equal to the experimentally determined simplified area shares (Figs. 3.8b–3.8e).
As it can be seen, e.g., in Fig. 3.7d, grains of very similar but not necessarily identical orientations are present in the cross-sections of the microspecimens. Therefore, the interfaces between, e.g., two $\langle 111\rangle$-grains of the texture are treated as GBs, nevertheless, in the simulations. The small variations in, e.g., the $\langle 111\rangle$-orientation of such grains in the experiments are, however, not explicitly accounted for in the modeling approach. This strict classification into $\langle 111\rangle$- and $\langle 100\rangle$-orientations seems justifiable for the two microwires with smaller diameters (see Section 3.6.2). For the two larger microwires, however, several crystal orientations in between these two orientations are present (see also

Section 3.6.2). Therefore, two distinct texture-cases are investigated for each of the two microwires with bigger diameter: the case of equal area shares for both orientations (Fig. 3.8a), and the case of considering non-equally distributed orientations in a simplified manner (Fig. 3.8d and Fig. 3.8e, respectively).

3.6.5 Material and model parameters

Regarding the material parameters, literature values are used for the anisotropic elastic constants of gold ($C_{1111} = 186\,\text{GPa}$, $C_{1122} = 157\,\text{GPa}$, $C_{1212} = 42$ GPa, see Rösler et al., 2006). In all simulations, the reference shear rate is $\dot{\gamma}_0 = 10^{-3}/\text{s}$, the rate sensitivity exponent is $p = 20$, and the drag stress is $\tau^{\text{D}} = 1$ MPa. An internal length scale, resulting from a first estimate of the observed dislocation density of the microwires, is $l_{\text{int}} = 1/\sqrt{\rho_0} \approx 3\,\mu\text{m}$ with $\rho_0 \approx 10^{11}/\text{m}^2$. The corresponding value for the normalization constant (see also Eq. (3.33)) is $g_0 = \sqrt{2W_0/K_G} = \sqrt{2W_0/(l_{\text{int}}^2 E_{\text{avg}})} = 0.2/\mu\text{m}$ with the average Young's Modulus $E_{\text{avg}} \approx 65$ GPa of all gold microwires (from Chen et al., 2015), the average shear modulus $G_{\text{avg}} \approx 24$ GPa for a Poisson's ratio of $\nu = 0.35$, and an initial defect energy density of $W_0 = \bar{\alpha} G_{\text{avg}} = 12$ GPa. Using this, resulting, value of g_0, however, leads to results substantially overestimating the experimental data in the elastic-plastic transition regime. Instead, the parameter g_0 is fitted to the experimental tensile curves by comparison of the slope in the elastic-plastic transition regime, resulting in $g_0 = 6.94/\mu\text{m}$. The internal length scale for this value is much smaller, $l_{\text{int,avg}} = \sqrt{2W_0/(g_0^2 E_{\text{avg}})} \approx 0.09\,\mu\text{m}$.

The remaining parameters are listed in the respective results sections. It is remarked that, at first, the same initial yield stress $\tau_0^{\text{C}} = 1$ MPa (e.g., Sachs and Weerts, 1930) is used for all slip systems in the investigated cases. This value also can be motivated by using the Taylor-hardening relation $\tau_0^{\text{C}} = \bar{\alpha} b G \sqrt{\rho_0}$ with Burgers vector magnitude $b = 0.224 \times 10^{-9}\text{m}$, and $\bar{\alpha} = 0.5$. Due to the additional yield criterion for the

GBs, the overall yield strength in the model results from the combined behavior of the bulk material and the GBs. Later on, the initial yield stress for the slip systems and the GB yield strength are scaled for the case of torsion loading (Section 3.6.6).
As it will be discussed in detail, in the next sections, the GB yield strength of the model cannot be used as a material constant anymore when modeling the mechanical response of the different microwires.

3.6.6 Simulation results

Influence of the two pre-dominant crystal-orientations on the size effects of the gradient plasticity simulations

Before the GP simulations are compared to experimental data, the basic influence by the $\langle 100 \rangle$-orientation and the $\langle 111 \rangle$-orientation on the overall mechanical response is discussed. The same material parameters from the preceding Section 3.6.5 are used, and an exemplary set of values is used for all remaining model parameters (see Table 3.7). Three different, simplified texture cases are investigated, in the following simulations. At first, all grains are assigned either the $\langle 100 \rangle$-orientation, or the $\langle 111 \rangle$-orientation. Then, equal numbers of grains are assigned the $\langle 100 \rangle$-orientation and the $\langle 111 \rangle$-orientation, respectively.

Table 3.7: Model parameters used for tensile test and torsion test simulations for different texture cases with $\langle 111 \rangle$- and $\langle 100 \rangle$-orientations.

D_{sim} (µm)	τ_0^{C} (MPa)	Ξ_0^{C} (Nm^{-1})	Θ (MPa)	τ_∞^{C} (MPa)
15	1	11	120	19
25	1	11	120	19
40	1	11	120	19
60	1	11	120	19

In all cases, the discretization of the GBs is "maintained" and a GB yield strength is assigned to the respective FE nodes. Thus, this approach allows to investigate the texture influence for identical GB properties within the model. Variations of the crystal orientations within the cross-sections, e.g., by a rotation around the loading axis, are not considered in this simplistic approach. In order to facilitate a comparison to the subsequent simulation results and the experimental data, the engineering stress is evaluated for tensile loading and the maximum shear stress is evaluated for torsion loading.

Considering grains to be exclusively either in the ⟨100⟩-orientation or the ⟨111⟩-orientation, the mechanical tensile responses depicted in Fig. 3.9a are obtained. A size effect is observable for both orientations.

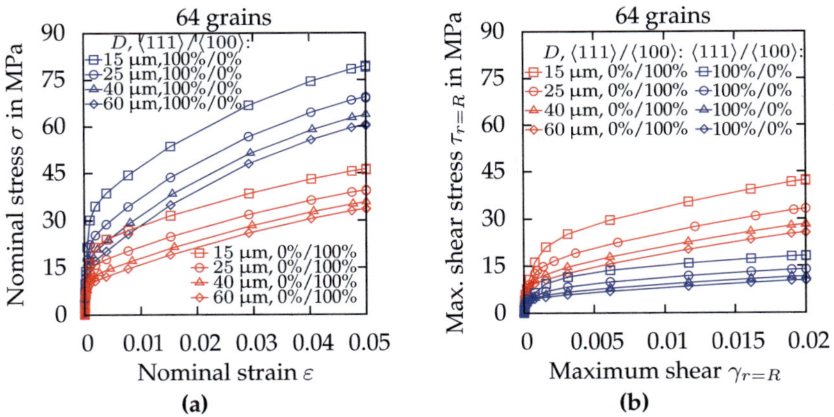

Figure 3.9: (a) Stress-strain curves of tensile-loaded simplified grain aggregates in comparison to (b) Maximum shear stress vs. maximum shear of the same aggregates under torsion loading for identical parameters. Either ⟨111⟩- or ⟨100⟩-orientations assigned to all grains of the individual aggregates.

The magnitude of the size effect is slightly more pronounced for the ⟨111⟩-orientation. For torsion loading, this effect is reversed (see Fig. 3.9b). In addition, it is noticeable that the ⟨111⟩-orientation responds

significantly stiffer for tensile loading than the ⟨100⟩-orientation (compare both cases in Fig. 3.9a). For torsion loading, this effect is reversed, too (see Fig. 3.9b).
Considering, however, an equal number of grains from both crystal-orientations, different mechanical responses under tensile loading and torsion loading are still obtained, which are depicted in Fig. 3.10.

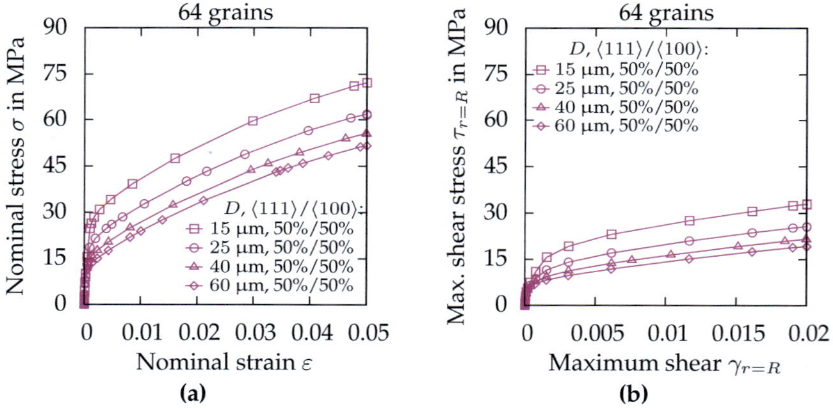

Figure 3.10: (a) Stress-strain curves of tensile-loaded simplified grain aggregates in comparison to (b) Maximum shear stress vs. maximum shear of the same aggregates under torsion loading with identical parameters. Approximately equal area shares of ⟨111⟩- and ⟨100⟩-orientations are used.

In comparison to Figs. 3.9a–3.9b, it is observable that the size-effect magnitudes as well as the overall yield strengths in the mechanical responses for the different load cases are still significantly affected by both crystal-orientations, although an equal number of grains from both orientations is assigned. In the following, this is investigated, in more detail, in comparison to experimental data.

Gradient plasticity simulations using an equal number of $\langle 100 \rangle$- and $\langle 111 \rangle$-oriented grains

In the following, the simulations results of the grain aggregates used to model the mechanical response of gold microwires under tensile and torsion loading are presented. At first, an equal number of $\langle 111 \rangle$- and $\langle 100 \rangle$-orientated grains is used, i.e., the diameter-dependent texture of the investigated specimens is not considered. The model parameters τ_∞^{C} (saturation stress of slip systems), Θ (initial hardening modulus), and Ξ_0^{C} (yield strength of the GBs) are fitted to the experimentally determined stress-strain curves (see Fig. 3.11a for experimental data and tensile simulations with the fitted parameters).

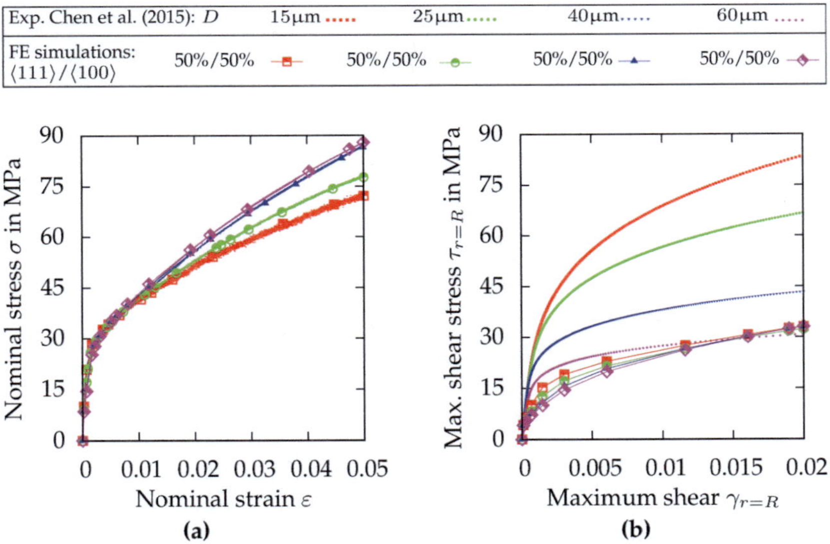

Figure 3.11: (a) Stress-strain curves of tensile-loaded simplified grain aggregates in comparison to experimentally observed oligocrystalline microwire responses by Chen et al. (2015), model parameters calibrated to the tensile experiment. (b) Maximum shear stress vs. maximum shear of the same aggregates under torsion loading compared to results by Chen et al. (2015), model parameters identical to (a). Equal cross-section area shares of $\langle 111 \rangle$- and $\langle 100 \rangle$-orientations used in the FE simulations. Figure reprinted from Bayerschen et al. (2016b) with permission from Springer.

Thereby, the parameter values, listed in Table 3.8, are obtained. Subsequently, using the same values, torsion test simulations are performed.

Table 3.8: Model parameters calibrated to tensile data for identical $\langle 111 \rangle$ and $\langle 100 \rangle$ area shares. Table reprinted from Bayerschen et al. (2016b) with permission from Springer.

D_{sim} (μm)	τ_0^{C} (MPa)	Ξ_0^{C} (Nm^{-1})	Θ (MPa)	τ_∞^{C} (MPa)
15	1	11	120	19
25	1	18	130	26
40	1	29	144	40
60	1	43	155	39

The maximum shear stress $\tau_{r=R} = 2M_T/(\Pi R^3)$ is plotted over the maximum shear $\gamma_{r=R}$ in Fig. 3.11b. No pronounced size effect is obtained in the simulations of the torsion loading with equal area shares of both orientations and this set of parameters. All grain aggregates exhibit similar overall mechanical responses for torsion loading.

Gradient plasticity simulations considering the simplified texture of the microwires

In order to investigate the influence of the microwire-specific texture, simulations considering the trend towards higher $\langle 100 \rangle$-oriented area shares for smaller wire diameters are carried out. Therefore, the experimental results of the individual textures obtained from the approach in Fig. 3.7 are used (see Table 3.5 for experimental data and Figs. 3.8b–3.8e for the grain aggregates). The model parameters are fitted to the experimental tensile response, again (see Fig. 3.12a). The resulting set of parameters is listed in Table 3.9. Using these parameters in the torsion test simulations, the results in Fig. 3.12b are obtained.

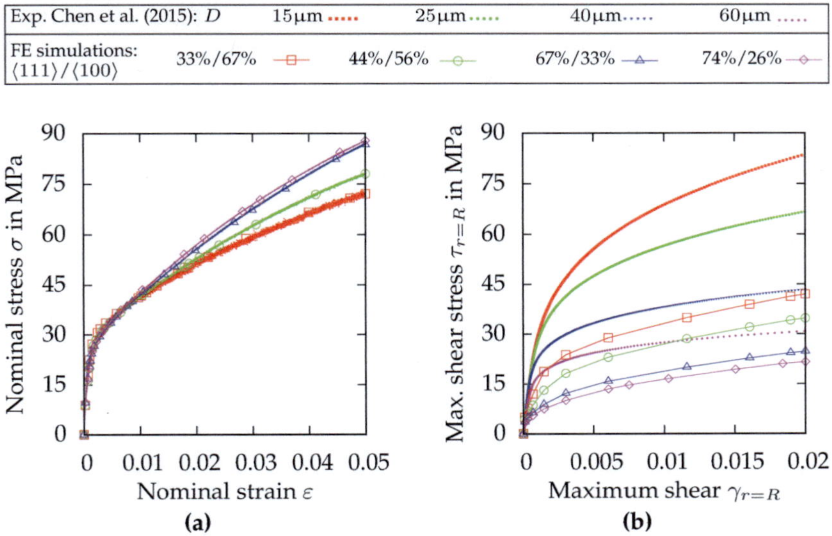

Figure 3.12: (a) Stress-strain curves of tensile-loaded simplified grain aggregates in comparison to oligocrystalline microwire experiments by Chen et al. (2015), model parameters calibrated to tensile experiment. (b) Maximum shear stress vs. maximum shear of the same aggregates and parameters for torsion loading compared to experiments by Chen et al. (2015). The experimentally determined, simplified cross-section area shares of $\langle 111 \rangle$- and $\langle 100 \rangle$-orientations are used in the FE simulations.

Table 3.9: Model parameters calibrated to tensile data using the experimentally determined, simplified cross-section area shares of $\langle 111 \rangle$- and $\langle 100 \rangle$-orientations. Table reprinted from Bayerschen et al. (2016b) with permission from Springer.

D_{sim} (µm)	τ_0^{C} (MPa)	Ξ_0^{C} (Nm^{-1})	Θ (MPa)	τ_∞^{C} (MPa)
15	1	13	143	21
25	1	19	135	28
40	1	25	143	37
60	1	37	145	35

A pronounced size effect is observable in the torsion test simulation results, although the experimental tensile responses, showing a small, inverse size effect, have been used to obtain the model parameters. The magnitude of the size-effect under torsion loading, however, is underestimated. It should be noted that the magnitude of the modeled size effect under torsion becomes smaller for increasing deviations of the texture of the two thicker microwires from the simplified approach taken for the grain aggregates (comparing the two curves of the thicker grain aggregates in Fig. 3.12 to the respective curves in Fig. 3.11).
The distributions of the accumulated plastic slip γ_{ac} for tensile and torsion loading, with and without consideration of the simplified microwire textures, are depicted in Figs. 3.13a–3.13b. Under both loading conditions, the plastic field distributions of the grain aggregates are very similar if the grain orientations are equally assigned.

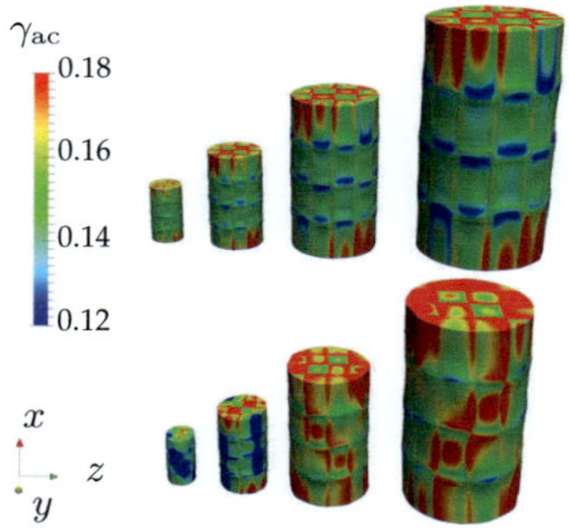

(a)

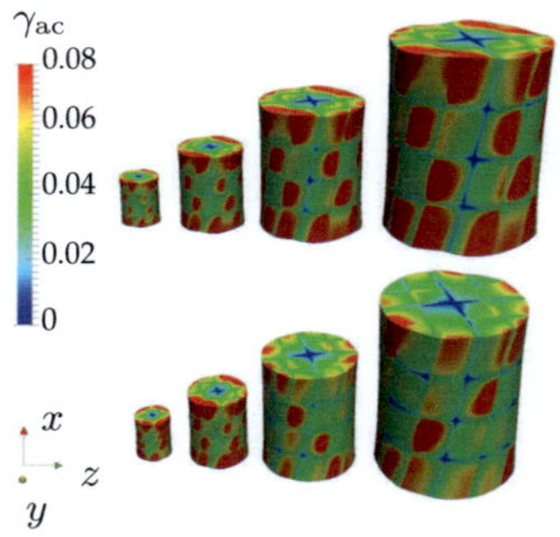

(b)

Figure 3.13: Field distributions of accumulated plastic slip for simplified grain aggregates using equal cross-section area shares of $\langle 111 \rangle$- and $\langle 100 \rangle$-orientations (top) and using experimentally determined, simplified cross-section area shares of $\langle 111 \rangle$- and $\langle 100 \rangle$-orientations (bottom): (a) tensile test simulations, (b) torsion test simulations. The deformation of all results has been geometrically scaled by a factor of five with regard to the displacement. Figure reprinted from Bayerschen et al. (2016b) with permission from Springer.

If the simplified microwire textures are considered, however, different field distributions of the accumulated plastic slip are obtained for the investigated aggregates of different diameters. It is remarked that the locally developing high values of γ_{ac} at the corners of the grains close to the GBs (see, for example, the top surface of the largest grain aggregate in Fig. 3.13b, bottom) result from the activation of favorably oriented slip systems in combination with the impediment of plasticity at the GB nodes due to the GB yield strength. In the ideally aligned $\langle 100 \rangle$-orientation case, the slip systems are activated such that a fourfold-symmetry of γ_{ac} is obtained for a single-crystalline torsion test simulation (see, e.g., Fig. 2.5). The spatial position of the maxima of γ_{ac} due to the activation of the slip systems coincides with the position of the observed localizations of γ_{ac}. Therefore, the observable higher values of γ_{ac} develop close to the corners of these grains. A similar localization would also occur if the GB, there, was modeled using a curved shape.

Gradient plasticity simulations considering the simplified texture of the microwires and scaling of parameters

In the experiments, an increase in the overall yield strength is observable, comparing the torsion test results to the tensile test results of the microwires. Therefore, the parameters used for torsion loading are scaled accordingly. It is investigated how this affects the size-effect prediction of the model. It appears that the initial overall yield strengths are significantly underestimated by keeping the initial yield stress of the slip systems as well as the GB yield strength identical under torsion loading (see Fig. 3.12b). Thus, these two model parameters (which constitute the overall yield strength, e.g., $R_{p0.2}$ in the model response) are altered. Therefore, the overall initial shear stresses are estimated based on the experimentally determined individual proof stresses $\tau_{0.2}^{\mathrm{te}}$ and $\tau_{0.2}^{\mathrm{to}}$, respectively (see Table 3.10).

Table 3.10: Yield strength scaling: maximum shear stresses at the onset of yielding in the experiments ([1]: from Chen et al., 2015) are used to determine scaling factors for the model parameters under torsion loading. $\tau^{\mathrm{te}}_{0.2}$ obtained by $\tau^{\mathrm{te}}_{0.2} = 0.5\sigma_{\mathrm{eng},0.2}$. Table reprinted from Bayerschen et al. (2016b) with permission from Springer.

Wire D_{sim} (μm)	Tensile shear stress[1] $\tau^{\mathrm{te}}_{0.2}$ (MPa)	Torsion shear stress[1] $\tau^{\mathrm{to}}_{r=R,0.2}$ (MPa)	Shear stress ratio $s = \tau^{\mathrm{to}}_{r=R,0.2}/\tau^{\mathrm{te}}_{0.2}$
15	15.8	54.3	3.4
25	16.4	39.5	2.4
40	14.4	31.7	2.2
60	13.9	22.7	1.6

The scaling factors from the initial overall yield strengths in the experiments under both loading conditions are used to scale the model parameters from the tensile loading to model parameters for the torsion loading. By this procedure the parameters in Table 3.11 are obtained.

Table 3.11: Torsion test simulation parameters, τ^{C}_0, Ξ^{C}_0 from the indicated Tables scaled by s from Table 3.10, $\tau^{\mathrm{C}}_{\infty,s}$ obtained by keeping $\Delta\tau^{\mathrm{C}} = \tau^{\mathrm{C}}_\infty - \tau^{\mathrm{C}}_0$ identical to the tensile test. Table reprinted from Bayerschen et al. (2016b) with permission from Springer.

D_{sim} (μm)	$s\tau^{\mathrm{C}}_0$ (MPa)	$s\Xi^{\mathrm{C}}_0$ (Nm^{-1})	Θ (MPa)	$\tau^{\mathrm{C}}_{\infty,s} = s\tau^{\mathrm{C}}_0 + \Delta\tau^{\mathrm{C}}$ (MPa)	Table
15	3.4	44	143	23.4	3.9
25	2.4	46	135	29.4	3.9
40	2.2	55	143	38.2	3.9
60	1.6	59	145	35.6	3.9
40	2.2	64	144	41.2	3.8
60	1.6	69	155	39.6	3.8

It is remarked that the saturation stresses τ^{C}_∞ of the slip systems are adjusted such that the hardening behavior of the microwires is kept the same, compared to the cases without scaling the initial yield stresses τ_0^{C}. At first, only the initial yield stress of the slip systems is scaled (see Table 3.11, second column) but the yield strength of the GBs is kept identical to the parameters obtained from the tensile loading (see Table 3.8 and Table 3.9). The torsion test simulation results obtained by this approach are depicted in Fig. 3.14a.

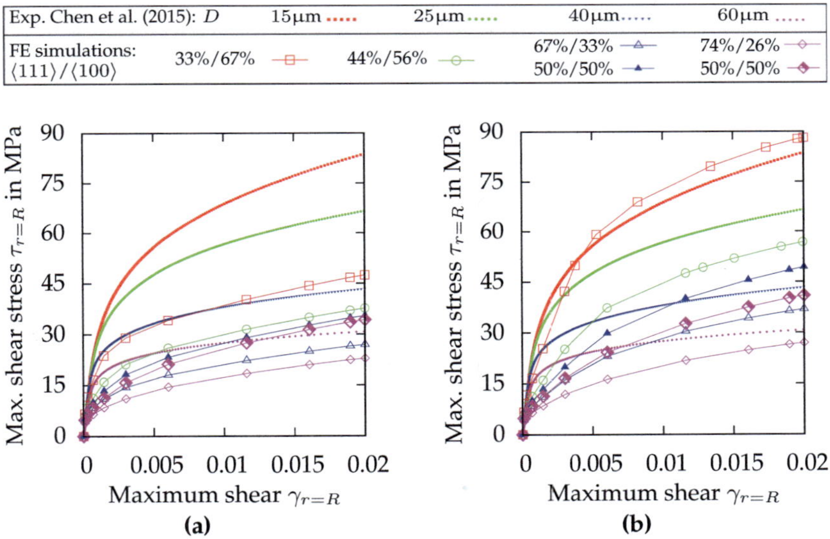

Figure 3.14: Maximum shear stress vs. maximum shear of torsion-loaded simplified grain aggregates in comparison to experimentally observed oligocrystalline microwire responses by Chen et al. (2015). (a) using scaled initial yield stress $s\tau_0^{\mathrm{C}}$ of the slip systems, (b) using scaled initial yield stress $s\tau_0^{\mathrm{C}}$ of the slip systems and scaled yield strength $s\Xi_0^{\mathrm{C}}$ of the GBs. For each microwire diameter, the experimentally determined, simplified cross-section area shares of ⟨111⟩- and ⟨100⟩-orientations are used in the FE simulations. For the two thicker microwires, additional FE simulations using scaled model parameters and equal cross-section area shares of ⟨111⟩- and ⟨100⟩-orientations are depicted. Figure reprinted from Bayerschen et al. (2016b) with permission from Springer.

Although the observed size effect magnitude in the simulations is slightly larger than the magnitude when using the previous initial yield stresses, the employed scaling of the slip system yield stresses is not sufficient to account for the size effect. For the two thicker microwires, the grain aggregates, using equal cross-section area shares of both orientations, show stiffer responses than the ones using the experimentally determined, simplified area shares of the orientations. The mechanical response of the microwire with $D = 60\ \mu\text{m}$ is captured better by using equal cross-section area shares of both orientations. However, for the microwire with $D = 40\ \mu\text{m}$, the mechanical response is still underestimated using equal area shares.

Since the yield strength of the GBs contributes to the overall yield strength of the grain aggregates in the simulations, this parameter is also scaled, subsequently (see Table 3.11, third column). The results are depicted in Fig. 3.14b. It is observable that the magnitude of the size effect is captured much better by scaling both the initial slip system yield stress and the GB yield strength. In addition, the mechanical responses of the simulations are much closer to the experimental results for all microwires. It is also notable that the experimental results of the thicker microwires are both in between the two respective simulated cases of the orientation distributions in the cross sections.

The field distributions of the accumulated plastic slip γ_{ac} of the results with scaled initial yield stress of the slip systems are depicted in Fig. 3.15a, and the distributions with additionally scaled GB yield strength are depicted in Fig. 3.15b. If only the initial slip system yield stress is scaled, the magnitude of the field distributions is only slightly reduced (see Fig. 3.15a compared to Fig. 3.13b). However, if the GB yield strength is additionally scaled, the GBs respond significantly stiffer and smaller values of γ_{ac} are observed, there (see arrows in Fig. 3.15b). This can be inspected more closely by evaluating the average γ_{ac} over circular line segments on the top surface (i.e., the surface at max. x-coordinate, see Fig. 3.15c). Comparing these averaged distributions for the first five time steps of the simulations shows that the scaling of the initial slip system yield stress slightly reduces the plastification in radial direction (Fig. 3.15d, see II compared to I).

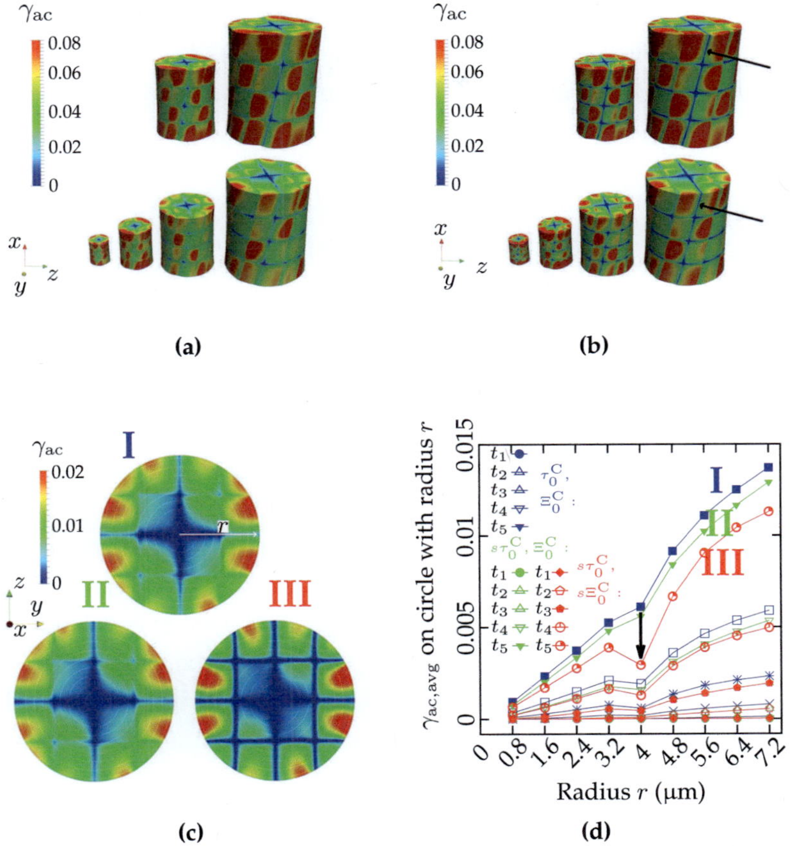

Figure 3.15: (a–b) Torsion test simulations (deformation geometrically scaled by a factor of five w.r.t. displacement): field distributions of accumulated plastic slip for simplified grain aggregates using equal cross-section area shares of ⟨111⟩- and ⟨100⟩-orientations (top) and using experimentally determined, simplified cross-section area shares of ⟨111⟩- and ⟨100⟩-orientations (bottom). (a) The initial yield strength of the slip systems has been scaled, (b) the initial yield strength of the slip systems and the grain boundary yield strength have been scaled. (c) Top surface of 15 μm-grain-aggregate at the end of the fifth time step for the cases of unscaled (I) and scaled parameters (II/III). The used line segments for obtaining the plots in (d) are visualized. (d) Average accumulated plastic slip evolution during the first five time steps on the corresponding circular line segments for the cases of unscaled (I), and scaled model parameters (II/III). Figures reprinted from Bayerschen et al. (2016b) with permission from Springer.

The scaling of the GB yield strength, however, leads to pronounced gradients in the γ_{ac}-distribution close to the GBs (see III in Fig. 3.15d and the arrow-indicator at $r = 4\,\mu m$, near the GB).

3.7 Discussion of the simulation results with focus on the size effects

Aggregates of grains with simplified shape and orientation distributions have been used in GP simulations to model the mechanical response of oligocrystalline microwires. While in the preceding comparison to experimental data from the literature (Section 3.5.3) the GB yield strength could be taken as a constant to reproduce the overall size-effect, this is not the case anymore in the second investigation presented (Section 3.6.6). The average grain size of the microwires is considered in the simulations but the variations of grain size for different grains within the cross-sections are not (Tables 3.5–3.6). Due to computational limitations, only aggregates up to 64 grains could be used. Therefore, in terms of the average grain sizes used in the modeling, smaller deviations remain for the microwires of two diameters, and larger deviations for the microwires of the other two diameters. Considering that the (non-destructive) experimental 3D-characterization of grain morphologies has made significant progress during recent years (see, e.g., Cabus et al., 2014; Hounkpati et al., 2014; Toda et al., 2016, and the review by Maire and Withers (2014)), it would be desirable to perform simulations using the actual morphology of all grains of the microwires, in order to have an advanced, more accurate representation of the microstructure in the simulations. For such an approach, the FE-meshes need to be substantially refined, for example, when modeling smaller grains of non-spheroidal or non-cuboid shape. This would require also further numerical improvement of the taken approach for the computational implementation.

Using equally distributed, ideally oriented, $\langle 100\rangle$- and $\langle 111\rangle$-grains, the experimentally observed tensile response of microwires by Chen et al. (2015) could be fitted using the GP model (Fig. 3.11a). The torsion response obtained in simulations with identical parameters and orientations as in the tensile response, however, shows a similar effect as the tensile response and does not match the classic size effect observed in the experiments for the torsion loading (Fig. 3.11b).
If the (simplified) texture of the microwires is considered in the grain aggregate simulations, the tensile response can be fitted, too (Fig. 3.12a). Then, however, a pronounced size effect is observable under torsion loading (Fig. 3.12b). The arising differences in the torsion responses regarding the texture can be attributed to the different orientations of slip systems. For the ideal $\langle 100\rangle$-orientation, more slip systems are favorably oriented for an activation under the investigated tensile loading compared to the ideal $\langle 111\rangle$-orientation. This behavior is reversed under torsion loading. Therefore, grain aggregates with a higher cross-section share of $\langle 100\rangle$-orientation respond significantly stiffer under torsion loading than grain aggregates with a higher $\langle 111\rangle$-orientation cross-section share (see also the discussion of the single-crystalline case in Section 2.5.4). It should be noted that the magnitude of the size effect in the model varies, depending on the chosen texture of the two larger aggregates (comparing, for example, the two curves of the thickest grain aggregate in Fig. 3.14a). Considering that, for instance, the largest microwire (see, e.g., Fig. 3.7f) shows larger area shares with higher misorientations from the two modeled ideal orientations, one would expect the texture-contribution to the size effect to be smaller than it is obtained by using the idealized texture from Fig. 3.8e.
The fitted model parameters are not unique for both cases. However, in parts, they show similar values and, overall, the same trends such as the increase in slip system saturation stress with increasing wire diameter (Tables 3.8–3.9). This increase saturates for the two largest microwires.
In the initial development of the present model, the GB yield strength

has been intended to be a material constant. However, using a unique value of the GB yield strength for all investigated microwire diameters is not applicable. On this account, it is also noted that the ratio of average grain size to wire diameter, in the experiments, is not unique for different wire diameters. For the simulated grain aggregates, however, the same ratio has been used for all discretizations. Additionally, in the experiments, the microstructure of the microwires has been annealed by different heat treatments for different wire diameters. More refined grain aggregates would be desirable to be used in the simulations but the arising computational costs appear to be prohibitive at the current time. From Tables 3.8–3.9 it is evident that different values of the GB yield strength Ξ_0^C have to be used in order to fit the experimental results. This is also due to the computational limitation regarding the simulated average grain sizes. If all experimental average grain sizes could be matched exactly in the simulations, a wire-diameter-independent value of the GB yield strength would be expected. The different values of Ξ_0^C, therefore, compensate also for the deviations in the modeling of the grain size. In order to visualize this influence, the values of Ξ_0^C are plotted in Fig. 3.16 over the difference in the ratio of average grain size d_{avg} to wire diameter D between the simulations and the experiments. Ideally, for the simulations, the ratio should be used that was determined in the experiments. If this was the case, all points in the plot should be on the dashed vertical line, and a wire-diameter-independent value of the GB yield strength is expected. Summarizing, it appears to be necessary to adapt the values of the GB parameter in the model to the specific microwires when the grain size cannot be modeled exactly.

Due to the fact that all considered microwires in this work exhibit comparably low shares of low-angle GBs (on average $< 10\%$ for angles $< 10°$), and that a model with a non-orientation dependent GB yield strength is used, effects of varying in-plane GB misorientations are not investigated (Bayerschen et al., 2016b). Instead, it has been focused on the effects stemming from the crystal-orientations.

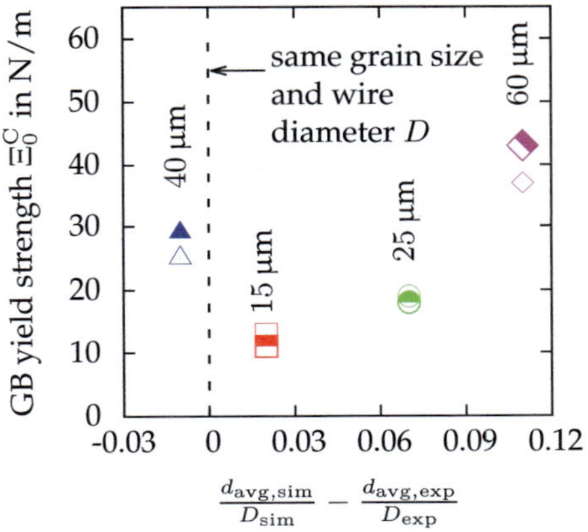

Figure 3.16: Grain boundary yield strength Ξ_0^C over the difference between the ratio "average grain size / wire diameter" of the simulations and the experiments. Values of Ξ_0^C from Tables 3.8–3.9. Grain sizes and wire diameters from Tables 3.5–3.6.

The activation of slip systems depends significantly on these, as it is shown by way of an example for the two distinct orientations considered in the modeling approach. The consideration of the simplified microwire textures enables to capture both the inverse size effect under tensile loading and the classic size effect under torsion loading. However, the magnitude of the size effect under torsion loading is underestimated (Fig. 3.12b). Scaling the initial yield stress of the slip systems by taking into account the experimentally determined 0.2-proof-stresses improves the simulation results, but not enough to obtain a good estimate of the size effect magnitude (Fig. 3.14a). If, additionally, the GB yield strength is scaled, a better approximation of the size effect magnitude is obtained (Fig. 3.14b). In parallel, the simulated accumulated plastic slip gradients close to the GBs are increased by this procedure (Fig. 3.15d). Such gradients in the model serve as an approximation of the inhomogeneity of the plastic deformation. The GND-densities calculated from the experiments show pronounced variations due to the microstructural characteristics (see, e.g., Fig. 4.30 in Chen (2013), however, for larger deformations than in the present work). Thus, the simulation results with larger gradients seem to be closer to the experimental observations. A more detailed assessment requires the determination of GND-densities in many cross-sections of the microwires for smaller overall deformations (than in Chen, 2013).

From a materials science point of view, it appears questionable that a different GB yield strength in the torsion test simulations results gives better agreement with the experiments than using the values as obtained from the tensile test fits. However, the initial distribution of dislocations has also not been considered in the simulations. Therefore, possibly differing activation stresses within the dislocation distributions of the same wires under different loading conditions are not modeled. These might be caused by the activation of dislocations close to the surface of the microwires of different thickness under torsion loading. Furthermore, the scaling of initial slip system stress and GB yield strength for the torsion loading simulations can only give an estimate of the mechanical response. The used model for GB yielding does not take into account the different orientations of the grains. In general, however, one would expect the GB yield strength to depend on the crystallographic orientations of the adjacent grains and of the GB itself, i.e., on the specific type of the GB. Several approaches to consider the crystallographic misorientation have been proposed in the literature, see Bayerschen et al. (2016a) for a recent review as well as Section 6.2.2 of the present work.

In the overall torsion response of simulations and experiments, deviations remain for the microwire with $D = 25\,\mu\mathrm{m}$ (Fig. 3.14b). These deviations could be reduced by employing a smaller grain size in the simulations that is closer to the experimentally determined average grain size. For the largest microwire, it would also be desirable to perform simulations with a refined grain size, although the influence of the grain size for the larger microwires is not expected to be as crucial as it is for smaller microwires.

The texture of the microwires has been considered in the simulations by a simplified approach (Fig. 3.7b). A refined approach would require the experimental characterization of many cross sections to obtain detailed results of the variations of texture with regard to the deviations from the ideal $\langle 100 \rangle$- and $\langle 111 \rangle$-orientations. The approach of a simplified texture characterization appears justifiable for the two smaller microwires (Fig. 3.7a and Fig. 3.7d) that show less cross-section shares with orientations deviating from the two considered ideal orientations. For the two thicker microwires (Figs. 3.7e–3.7f), however, the examined cross-sections show higher area shares with orientations deviating from the ideal orientations investigated. Therefore, two texture cases (for $D = 40\,\mu\text{m}$ with the aggregates from Fig. 3.8a and Fig. 3.8d, for $D = 60\,\mu\text{m}$ with the aggregates from Fig. 3.8a and Fig. 3.8e, respectively) have been used in the simulations to obtain an estimate of the mechanical responses. The mechanical response of the microwires in the experiment is in between the responses of the two simulated texture cases (Figs. 3.14a–3.14b) for both diameters. Thus, the actual texture of the microwires could also be in between the two idealized cases simulated.

Regarding the employed gradient plasticity model, several simplifications are addressed, in the following. First of all, an accumulated plastic slip and its gradient are used instead of considering all plastic slips and their gradients. Physically richer models incorporate all plastic slips or dislocation densities, e.g., Gurtin et al. (2007); Bardella et al. (2013), as additional DOFs. Using the accumulated plastic slip and its gradient leads to a reduction of the defect stress modeling since only one higher-order stress, work-conjugate to the gradient of γ_{ac}, is modeled. Considering all gradients, instead, would allow to include backstresses related to the individual slip systems but the model merit of manageable computational times would possibly be lost. In addition, it has been assumed that the accumulated plastic slip measure is continuous across GBs. The plastic slips of individual slip systems are, however, not necessarily continuous. Although this simplification could be justifiable in certain cases, it is not expected to be of general validity (see also Bayerschen et al., 2015).

Furthermore, the accumulated plastic slip has been set to vanish throughout the simulation volume, in the beginning of the simulation. This appears to be reasonable since the annealing process produces oligocrystalline microwires with comparably low initial dislocation density content. It would, however, be desirable to consider the initial dislocation density distribution. Detailed experimental data of the dislocation density distributions within all grains and in many cross sections would need to be obtained for meaningful non-zero choices of the initial accumulated plastic slip distribution. The effort for this appears to be prohibitive at the current time.

As it is also commonly done in the literature, the defect energy, related to the gradients in the accumulated plastic slip, has been taken to be of quadratic form. However, other approaches (e.g., linear) have been proposed in the literature as well. Different accumulated plastic slip distributions would be obtained with a non-quadratic defect energy. This includes steeper gradients close to GBs for, e.g., a linear defect energy compared to a quadratic one. The defect energy also influences the magnitude of size effects (see Chapter 5). Therefore, it could be interesting to additionally consider the influence of the defect energy formulation in future works on size effects. A non-quadratic generalization of the defect energy formulation is, therefore, discussed in Chapter 5 including first simulations.

The parameter g_0 of the defect energy has been fitted to the experimental tensile responses. Since a simplistic quadratic defect energy is used in the current model, a direct connection to the microstructural internal length is not necessarily given. Recent works suggest that the free energy of dislocations should in fact be non-quadratic, e.g., Kooiman et al. (2015; 2016). Although the internal length scale is, in general, not expected to be constant, the accordance with the experimental tensile test results is given when using the same value of the defect energy parameter for all grain aggregates in the present approach. From a physical point of view, the internal length scale is determined by the microstructural characteristics of the specimen such as grain size, specimen size, dislocation spacing and dislocation source length (see, e.g., Zhang et al., 2014). The obtained value for the internal length scale of $l_{\mathrm{int,avg}} = 0.1$ µm is, however, at least of the same order as, for example, the value of the internal length scale of the gradient plasticity model by Aifantis and Willis (2004; 2005) which was determined in nanoindentation studies close to a GB (Aifantis et al., 2006). There, the interpretation was given that this length was the distance over which 90% of the dislocations are piled up.

The used model allows only to reproduce a limited size-effect magnitude of the torsional response of the gold microwires of different thicknesses based on the parameters from the related tensile tests. However, it is evident that, besides grain size, also textural features have to be taken into account when modeling the mechanical behavior of small structures under different loading conditions. In the present case it is demonstrated, that the size effect in torsion, determined in Chen et al. (2015), is significantly affected by the textural differences. If all the wires would possess similar textures, the differences in strength of the microwires with smaller diameter, compared to the ones with bigger diameter, would be less pronounced under torsion loading. In fact, the response would be expected to be similar for both loading cases. However, in many experimental works on size effects of microspecimens, the influence of the texture is often neglected (e.g., Fleck et al., 1994; Liu et al., 2012; 2013). As it is shown in the preceding chapter, the texture can have a significant influence on the differences in the overall mechanical responses and, therefore, it should be considered explicitly in future experimental characterizations as well as in simulation studies.

Chapter 4

Extension of the gradient crystal plasticity theory with grain boundary yielding by grain boundary hardening

4.1 Motivation

As it is shown in the preceding Chapter 3, the consideration of the resistance of GBs against plastic flow in the GP approach allows to model phenomena such as the overall size-effects on oligocrystalline microspecimen observed in experiments. However, a direct comparison of the numerical GP results obtained for local quantities, such as the distribution of plastic strain, to experimental data is not performed due to a lack of availability of such data in the used experimental reference works. Nevertheless, the prediction of plastic strain distributions with the GP model is of central interest, especially close to the GBs due to the arising gradients in the plastic field distributions, there. Therefore, an additional comparison of GP results to discrete dislocation dynamics (DDD) results is performed. The physically detailed modeling of dislocations and their interactions in, e.g., (the here used) DDD-codes by Weygand et al. (2002; 2009); Šiška et al. (2009), and others, allows to use such results as a data basis for comparison of GP results.

Due to the discrete character of DDD simulations, the results have to be averaged. This averaging is performed over many simulations. The DDD simulations presented in this chapter were carried out and post-processed by M. Stricker and D. Weygand (Institute of Applied Materials, KIT). The GP and DDD results shown in this chapter are published in Bayerschen et al. (2015), and the content of this chapter is largely taken from this work. At first, however, the DDD simulations are briefly summarized.

4.2 Summary of discrete dislocation dynamics simulations

4.2.1 Simulation setup, geometry and boundary conditions

A tricrystal composed of three cubic grains with edge length 0.75 µm is modeled under tensile loading (Fig. 4.1b). The GBs in the DDD model are impenetrable to dislocation movement, i.e., dislocations cannot pass the interfaces between the grains. Dislocations on the adjacent sides of the GBs can interact, nevertheless. This means that the elastic interactions of dislocations across GBs is preserved. Therefore, when the dislocation density is sufficiently large enough (like in the present study), this interaction of dislocations can be interpreted as an effective transmission of dislocations (Quek et al., 2014; Stricker et al., 2016). To each grain, an individual crystallographic orientation is assigned. Initially, all three grains are in $\langle 100 \rangle$-crystal-orientation with respect to the depicted coordinate system of Fig. 4.1a. In order to investigate different dislocation interaction behavior across the GBs in the model a misorientation angle φ around the tensile axis is imposed on the central grain (Fig. 4.1a). Strain-rate controlled tensile test simulations are carried out along the x-axis of the tricrystal geometry (cf. Fig. 4.1) with a strain

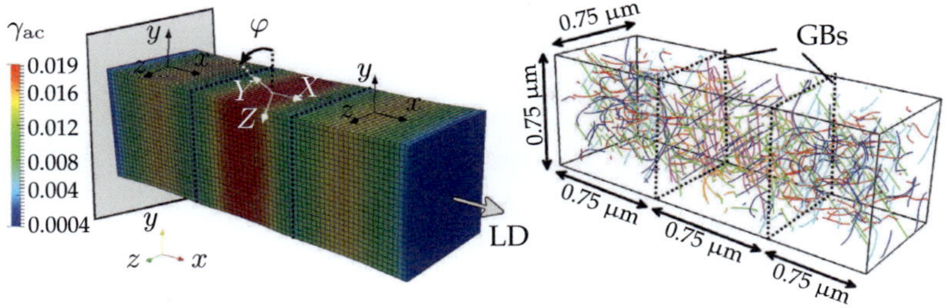

(a) (b)

Figure 4.1: (a) Accumulated plastic slip field distribution obtained in a gradient plasticity FEM simulation without lateral contraction at the boundaries and finite values of the GB yield strength (depicted are the Gauss-point subvolumes). (b) DDD simulation with impenetrable GBs (dashed lines indicate the GB planes). LD: loading direction. Basic figure from Bayerschen et al. (2015).

rate of $\dot{\varepsilon} = 5000\,\mathrm{s}^{-1}$ during the simulation time t. This value is chosen due to the small time scale at which DDD operates and it is assumed that the plastic behavior is independent of the strain rate (Senger et al., 2008). On the two boundary planes (in x-direction) the displacements $u_x(x = 0) = 0$ and $u_x(x = x_{\max}) = x_{\max}\dot{\varepsilon}t$, respectively, are prescribed. The displacements orthogonal to the loading axis are set to zero (except for one special case, cf. Section 4.3.4). All other boundaries are set to be traction-free and dislocations are allowed to leave the volume, there.

4.2.2 Material model

Isotropic linear elasticity is used for the FCC crystal-system. The elastic constants of aluminum are used (shear modulus $G = 27\,\mathrm{GPa}$, Poisson's ratio $\nu = 0.347$). Frank-Read sources are distributed randomly with respect to their positions and orientations in the grains. This assignment is performed under the restriction that each slip system in each grain is containing the same number of sources. The initial source length is chosen to vary between $0.16 - 0.27$ µm. These choices reduce artificiality which would be introduced into the model by a uniform source length.

The resulting initial dislocation density is about $\rho \approx 7.5 \times 10^{13}\,\mathrm{m}^{-2}$. All sources are approximately two times larger than the mean dislocation spacing $\ell = 1/\sqrt{\rho} \approx 0.12\,\mu\mathrm{m}$. This places the behavior of the dislocation system in the multiplication controlled plasticity regime (e.g., Kraft et al., 2010; Zhang et al., 2014). As a consequence, the hardening of the dislocation system is mainly influenced by dislocation reactions, and not by single-source controlled plasticity. Therefore, a comparison of results from the continuum GP model to the discrete DDD results can be performed.

4.2.3 Averaging procedure

Due to the discrete nature of DDD simulations, scatter occurs in the results. These depend on the dislocation structure and on the dislocation density. The higher the dislocation density, the closer are DDD results from individual simulations to a continuum-like result. The size of the setups considered in the present work, however, is within the size-effect regime. In this regime, individual dislocations control plasticity. Therefore, a suitable averaging procedure is necessary in order to obtain results comparable to continuum results. Therefore, an averaging over several simulations is performed for all investigated cases. Once all DDD simulations are carried out, the plastic strain of each realization is evaluated from the swept areas of all discrete dislocations. In this postprocessing procedure, slices perpendicular to the tensile axis are considered. Finally, the DDD results are averaged, for one set of BCs and crystal orientations over M simulations. The averaging procedure is described in Table 4.1. For the evaluation of plastic strain along the tensile axis, the number of slices is chosen to be 150, resulting in a slice thickness of $15\,\mathrm{nm}$. This choice is based on the evaluation of the mean dislocation spacing in the developing pile-ups close to the GBs which is approximately $1/\sqrt{\rho} \approx 25\,\mathrm{nm}$. Using 150 slices, the evaluation resolution is nearly two times higher than the mean dislocation spacing in the pile-ups. This resolution is sufficiently high to capture the strain gradients. The GBs, themselves, do not exhibit plastic strain but the slices adjacent to them do. Due to the performed averaging, the spatial distribution of plastic strain is rendered to be one-dimensional. All dislocations add to the produced plastic strain, regardless of their individual positions within the cross-sections. The averaged DDD results are used to calibrate the GP model parameters. Subsequently, the simulation results in terms of the plastic strain profiles can be compared for both models.

Table 4.1: Averaging procedure for the discrete dislocation dynamics simulations, developed and evaluated by M. Stricker and D. Weygand (Institute of Applied Materials, KIT). Procedure reprinted from Bayerschen et al. (2015).

1. The contribution of a slip system α to the plastic strain tensor is evaluated in each slice of the volume V via $\varepsilon^{\mathrm{p}}_{\alpha} = bA_{\alpha}/(2V)\,(\boldsymbol{d}_{\alpha} \otimes \boldsymbol{n}_{\alpha} + \boldsymbol{n}_{\alpha} \otimes \boldsymbol{d}_{\alpha})$, where b is the length of the Burgers vector, and A_{α} is the swept area of dislocations from a slip system α.
2. The slip system contributions $\varepsilon^{\mathrm{p}}_{\alpha}$ to the plastic strain tensor ε^{p} are superposed within each slice.
3. In each slice, the mean plastic strain is obtained by averaging the simulation results twice:
   a) Averaging the plastic strain arithmetically over $\sim$ 5 DDD simulations.
   b) Averaging the plastic strain over equivalent volumes with respect to the crystallographic mirror-symmetry in x-direction at $x = 0.5\,x_{\mathrm{max}}$. Thus, the data from DDD is effectively doubled and, therefore, further smoothed. This procedure is formally similar to the averaging procedure in Aifantis et al. (2009). There, however, it is averaged over all symmetry-equivalent components of the dislocation density tensor –instead of the plastic strain– and over all symmetry-equivalent positions along the x-axis.

4.3 Gradient plasticity simulations

4.3.1 Geometry, boundary conditions and crystal orientations

In this section, the setup of the FE simulations for the GP model is discussed (for details on the FE implementation see Wulfinghoff et al., 2013; Wulfinghoff and Böhlke, 2013, and Section 3.4). The tricrystal composed of 0.75 µm-wide cubic grains is loaded by Dirichlet conditions ($\Delta u_x = 0.005 L_0$, $L_0 = 2.25$ µm, cf. Fig. 4.1). Lateral contraction on the boundary planes at $x = 0$ and at $x = x_{\max}$, respectively, is prohibited (except for one special case, cf. Section 4.3.4). The restriction of lateral contraction (see also Fig. 4.1) is abbreviated by NLC ("no lateral contraction"), in the following. At the beginning of the simulation, the accumulated plastic slip γ_{ac} (and its micromorphic counterpart ζ) are set to zero throughout the simulation volume. Finite element nodes on the GB planes as well as on the boundary planes are assigned a GB yield strength and are set to microhard behavior, at the beginning. Once the (GB) yield condition at these nodes is fulfilled, plastic activity is allowed for, there. The motivation for assigning a GB yield strength to the boundary planes, too, is drawn from the BCs in the DDD simulations that lead to pile-ups of dislocations at the boundaries as well (cf. Section 4.2.2). For details on the active-set search for the (grain) boundary nodes in the GP simulations it is referred to Wulfinghoff et al. (2013) and Section 3.4.2.
An elastically isotropic, but plastically anisotropic aluminum-like material is considered throughout the following GP simulations. As a first approach to investigate the interaction behavior across GBs with both models, three representative cases for the misorientation of the central grain, $\varphi \in \{0°, 5°, 35°\}$, are selected. Thus, the ideal case of vanishing mismatch between the slip systems of adjacent grains (and, consequently, unrestricted interaction of dislocations across the GBs in the DDD setting) is supplemented by a case of small mismatch (yet

still strong dislocation interaction) and a case of large mismatch (weak dislocation interaction) between adjacent grains.
The chosen FE mesh for the GP simulations consists of $12 \times 12 \times 12$ elements for each grain, i.e., it features 25012 DOFs in total. This mesh is chosen as a compromise between computational time and accuracy. Compared to the chosen discretization, a refinement of the mesh, using approximately twice the number of DOFs, yields a relative error in the stress-strain response at the final time-step of less than 0.01.

4.3.2 Gradient plasticity model with Voce-hardening

Model parameters

The GP model parameters are calibrated such that the overall mechanical response matches the averaged stress-strain curves of the DDD simulations. This calibration is, in general, not sufficient by its own since distributions of the plastic strain, e.g., along a line segment of the whole volume, are not necessarily predicted correctly. Therefore, the local distribution of plastic strain has to be taken into account in the calibration, too. While, in principle, the determination of the GP model parameters in the fitting procedure is not unique, some guidelines, therefore, are shown in Table 4.2 which should help in the calibration. Regarding the elastic constants, the calibration yields a uniform Young's modulus of $E = 65$ GPa for all GP simulations (except for one special case, cf. Section 4.3.4). The value of the Young's modulus is slightly lower, compared to the value of 72.7 GPa used in the DDD model. This is due to the bow-out of dislocations from the very beginning of the loading (Bayerschen et al., 2015). The Poisson's ratio is kept identical to the DDD simulations ($\nu = 0.347$) and the resulting initial defect energy density is $W_0 = 0.5G = 12.06$ GPa.

Table 4.2: Fitting procedure for gradient plasticity model with Voce-hardening, reprinted from Bayerschen et al. (2015).

1. A least-squares fit (LSF) of the DDD stress-strain curves is obtained.
2. The cross-section averaged plastic strain profiles along the loading axis are obtained. This is performed in order to ensure comparability with the (averaged) plastic strain profiles from the DDD simulations. The occurring differences within the cross-section distributions of GP results are relatively small and not as pronounced as they are in the DDD simulations.
3. The Young's modulus of the GP model is calibrated in order to match the elastic stiffness obtained from the LSF of the DDD data.
4. The initial yield stress τ_0^{C} of the GP bulk model and the initial yield strength Ξ_0^{C} of the GBs as well as the initial yield strength $\Xi_{0,\partial\mathcal{B}}^{\mathrm{C}}$ of the boundary (planes at $x = 0$ and $x = x_{\max}$) are calibrated using the DDD-LSF and the averaged plastic strain profiles. Therefore, the plastic strain profiles along the loading axis of GP results and DDD results are compared. On that account, plastic strain profiles are obtained at three, representatively chosen, fixed overall plastic strain values in the well established plastic regime. The values used are $\varepsilon^{\mathrm{p}} \in \{0.001, 0.002, 0.003\}$.
5. In case of Voce hardening: The initial hardening modulus Θ and the saturation stress τ_∞^{C} are adjusted to the hardening behavior of the DDD-LSF.

It is remarked that the defect energy normalization constant g_0, although in principle introducing an internal length scale into the model, mainly controls the elastic-plastic transition behavior if the GP model with GB yielding is used (see Section 3.5.1). Consequently, a value of $g_0 = 16.95/\mu\text{m}$ is chosen such that the GP simulations show similar stress-strain results as the DDD simulations in the elastic-plastic transition regime. Hardening resulting from the defect energy is negligible compared to the hardening relations investigated in the following (cf. also the parameter study in Fig. G.1). The internal length scale in the GP model can be obtained from the normalization constant g_0 by $l_{\text{int}} = \sqrt{2W_0/(g_0^2 E)} \approx 36$ nm, considering a Young's modulus of $E = 65$ GPa. This result is, remarkably, of the same order of magnitude as the mean dislocation spacing in the pile-ups at the GBs of the DDD simulations (cf. Section 4.2.3). However, it is not expected that g_0 (and thus, the internal length scale) is a constant, in general. For all simulations, a reference shear rate of $\dot{\gamma}_0 = 10^{-3}/\text{s}$, a rate sensitivity exponent of $p = 20$, and a drag stress of $\tau^{\text{D}} = 1$ MPa are considered. The used penalty parameter is $H_\chi = 10^8$ MPa. At first, two cases of different misorientation are investigated. They are summarized in Table 4.3. In the case NLC35V, the misorientation of the central grain is 35° and, in the case NLC5V, the misorientation is 5°. The listed model parameters are obtained by the fitting procedure outlined in Table 4.2.

Table 4.3: Setup and model parameters of GP simulations for comparison to DDD results. The abbreviation NLC indicates that lateral contraction is prevented on the boundary planes at $x = 0$ and $x = x_{\text{max}}$. The gradient hardening contribution is negligible in the investigated NLC cases, see Fig. G.1 in the appendix. Data reprinted from Bayerschen et al. (2015).

Name	Angle φ	Hardening	$\Xi_{0,\Gamma}^{\text{C}}$ (N/m)	$\Xi_{0,\partial\mathcal{B}}^{\text{C}}$ (N/m)	τ_0^{C} (MPa)	τ_∞^{C} (MPa)	Θ (MPa)
NLC5V	5°	Voce	3.5	25	30.0	108.51	1075
NLC35V	35°	Voce	3.5	25	30.0	108.51	1075

Numerical results

At first, the framework presented in Chapter 3 is used taking into account the parameter calibration guidelines described above. A misorientation of 35° of the central grain is considered. This case exhibits weak interaction of dislocations across the GBs in the discrete simulations due to the substantially differing slip system orientations of the adjacent grains. The plastic strain profiles of the DDD simulations are evaluated and averaged at three constant overall plastic strains (cf. Fig. 4.2a). Subsequently, the cross-section averaged plastic strain profiles of the GP simulations are compared to the DDD profiles, see Fig. 4.2b. The model parameters for this case can be found in Table 4.3. When calibrating the GP parameters, it can be seen, e.g., in Fig. 4.2b, that the plastic strain profiles for $\varepsilon^{\mathrm{p}} = 0.001$ are in good agreement. The subsequent evolution of plastic strain close to the GBs, however, cannot be accounted for by the GP simulations using Voce hardening. Significant deviations occur (see exemplary red arrow indicators in Fig. 4.2b) which are caused by an obvious limitation of this approach to account for the accumulation of dislocations at the GBs observed in the discrete simulations. Next, for the case of a small misorientation of the central grain (5°), the stress-strain response of the GP model is fitted to the DDD results (Fig. 4.2c). The corresponding GP plastic strain profiles are in better agreement with the DDD simulation results within the grains (Fig. 4.2d), compared to the case of 35° misorientation (Fig. 4.2b). For small misorientations, the interaction of dislocations in the discrete simulations is higher, leading to a more homogeneous distribution of plastic strain over all three grains. However, similar deviations occur between GP and DDD results close to the GBs (Fig. 4.2d) as the evolution of the plastic strain and its gradients cannot be accounted for sufficiently by the used model. Therefore, an additional hardening relation for the GBs is investigated in the following and, after a calibration of the new hardening parameter, the GP results are compared to the DDD results, again.

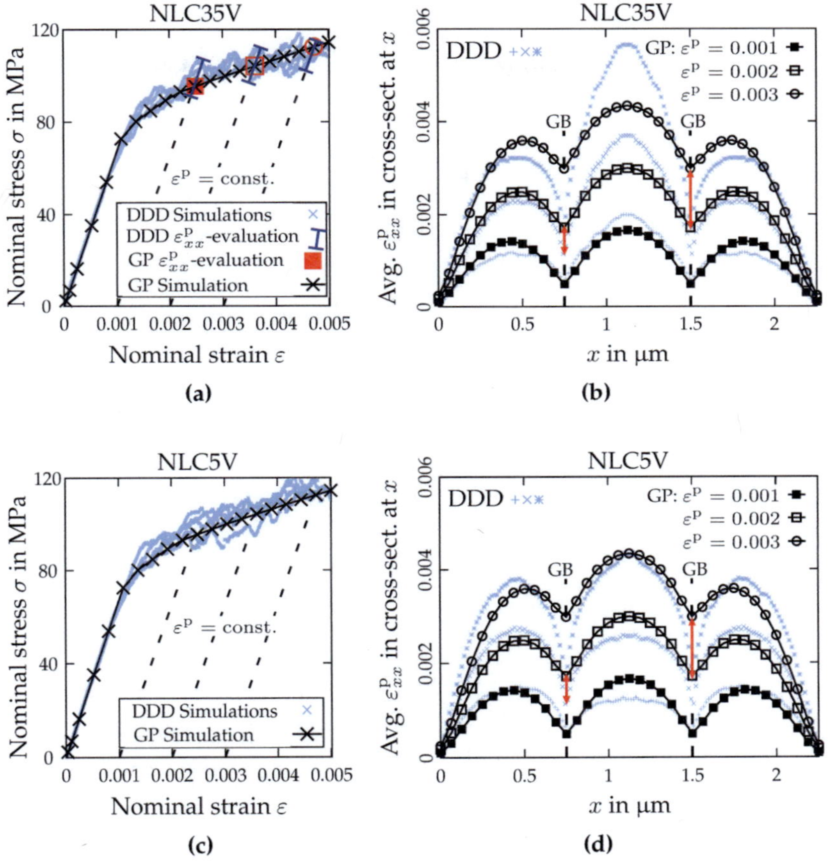

Figure 4.2: (a) Stress-strain curves of DDD and GP simulations for 35° misorientation. Blue error bars indicate that the DDD results are averaged at the corresponding overall plastic strains. (b) Distribution of cross-section averaged plastic strain along loading direction of DDD and GP simulations for 35° misorientation, obtained at the fixed overall plastic strain values from (a). (c,d) GP and DDD simulation results for 5° misorientation. Red arrows indicate deviations at the GBs. GP simulations obtained with quadratic defect energy. Central grain rotated by misorientation angle around loading axis. DDD data (a-d) and GP data (a,b) from Bayerschen et al. (2015).

4.3.3 Gradient plasticity model with grain boundary hardening

Motivation

The presence of GBs in micro-structured materials leads to dislocation pile-ups which influence the overall hardening behavior. In many continuum models, the dislocation-induced hardening mechanisms are modeled by, e.g., isotropic hardening relations for the bulk material such as the Voce-relation from the preceding chapter. When the overall mechanical properties are focused on, this type of modeling can be sufficient (see, e.g., Wulfinghoff et al., 2013). As it has been shown in Section 4.3.2, however, the evolution of the distribution of plastic strain is not captured by this approach. Therefore, it is investigated in the following, if the localization of plastic strain, resulting from pile-ups of dislocations observed at the impenetrable GBs in DDD simulations, can be accounted for by a GP model with GB yielding and GB hardening. A direct translation of the DDD grain-boundary conditions to the GP model is not possible due to the coarsening in the continuum approach. In order to account for hardening, as an alternative to the preceding bulk-hardening model, GB hardening is introduced. The GB hardening relation takes into account the accumulated plastic slip at the GBs. This explicit association of hardening to the GBs is motivated by observations from the DDD simulations. There, dislocations entangle localized as pile-ups close to the GBs. Within the context of GB yielding, it is also noteworthy that, in Aifantis et al. (2006), GB yield stresses are estimated based on indentation studies. The authors attribute the observed increase in hardness near the GB (and, thus, an increase in resolved shear stress) to dislocation pile-ups (Soer et al., 2005). Keeping this in mind, the explicit consideration of the GBs in the hardening relation of the GP model seems justifiable.

Extension of the grain boundary yield condition by grain boundary hardening

The basic GP framework from Chapter 3 is considered but the Voce-hardening relation of the bulk is neglected, in the following. Thus, $\beta = 0$, and the flow rule of the bulk is given by

$$\dot{\lambda}_\alpha = \dot{\gamma}_0 \left\langle \frac{\tau_\alpha^{\mathrm{d}} - \tau_0^{\mathrm{C}}}{\tau^{\mathrm{D}}} \right\rangle^p = \dot{\gamma}_0 \left\langle \frac{\tau_\alpha + \mathrm{div}\,(\boldsymbol{\xi}) - \tau_0^{\mathrm{C}}}{\tau^{\mathrm{D}}} \right\rangle^p . \qquad (4.1)$$

Consequently, hardening of the bulk is only induced by the gradient stress $\boldsymbol{\xi}$. For the following investigations, the gradient hardening contribution, however, is negligible as the hardening contribution of the GBs is much more pronounced.

In order to model the hardening of the GBs, the energy density (3.10) is extended by an additional quadratic term and, therefore, reads

$$W_\Gamma(\zeta) = \Xi_0^{\mathrm{C}} \zeta + \frac{1}{2} K_{\mathrm{H}} \zeta^2 . \qquad (4.2)$$

It is remarked that this energy approach requires an update of the stiffness matrix, see 1.b) in Table 3.2 for details.

Here, K_{H} is a parameter that describes the intensity of hardening of the GBs. The simple quadratic form of the additional energy density contribution is chosen as a first approach. Thereby, a hardening relation is obtained for the GBs that is linear in the micromorphic variable. However, other approaches could be feasible, too. The combination of (4.2) with the GB yield condition (3.19) yields

$$f_\Gamma = [\![\boldsymbol{\xi}]\!] \cdot \boldsymbol{n} - \Xi_\Gamma^{\mathrm{e}} = [\![\boldsymbol{\xi}]\!] \cdot \boldsymbol{n} - (\Xi_0^{\mathrm{C}} + K_{\mathrm{H}} \zeta). \qquad (4.3)$$

Here, it is clearly exhibited that the energetic GB traction Ξ_Γ^{e} is the sum of a constant initial value Ξ_0^{C} of the GB yield strength and a hardening term linearly dependent on the micromorphic variable ζ.

Model parameters

When GB hardening is used instead of Voce-hardening, the fitting procedure for the model parameters is slightly different, see Table 4.4. The steps 1.-4. are identical to the steps in Table 4.2. However, step 5. from Table 4.2 is substituted by fitting the new hardening parameter K_{H} of the GBs in the GP model to the DDD results (see Table 4.4).

Table 4.4: Fitting procedure for gradient plasticity model with grain boundary hardening, reprinted from Bayerschen et al. (2015).

1.-4. See Table 4.2

5. In case of GB hardening: The GB hardening parameter K_{H} is adjusted to give good agreement with the hardening behavior of the DDD-LSF. Additionally, the evolution of the plastic strain at the GBs is taken into account, i.e., the GP plastic strain profiles are compared to the (averaged) DDD plastic strain profiles at all three overall plastic strain values ε^{P}.

Considering the same tricrystals and crystal orientations from Section 4.3.2, the procedure leads to slightly different initial yield stresses of the slip systems and a slightly different GB yield strength (cf. values in Table 4.5 to Table 4.3).

Table 4.5: Setups and model parameters of GP simulations for comparison to DDD results. The abbreviation NLC indicates that lateral contraction is prevented on the boundary planes at $x = 0$ and $x = x_{\mathrm{max}}$. The gradient hardening contribution is negligible in the investigated NLC cases, see Fig. G.1. Data reprinted from Bayerschen et al. (2015).

Name	Angle φ	Hardening	$\Xi^{\mathrm{C}}_{0,\Gamma}$ (N/m)	$\Xi^{\mathrm{C}}_{0,\partial\mathcal{B}}$ (N/m)	K_{H} (N/m)	τ^{C}_0 (MPa)
NLC5G	5°	GB	1.5	25	1.8×10^3	33.5
NLC35G	35°	GB	1.5	25	1.8×10^3	33.5

Results with grain boundary hardening

Using the calibrated parameters from Table 4.5, the stress-strain responses depicted in Fig. 4.3a and Fig. 4.3c are obtained. The evaluation of the corresponding strain profiles (Fig. 4.3b and Fig. 4.3d) shows that, by using the GB hardening relation, much better agreement of the GP results with the DDD results is achieved. It can be seen (Fig. 4.3b and Fig. 4.3d), however, that, depending on the chosen rotation angle φ of the central grain, the strain profiles of the GP simulations are in better agreement with the DDD profiles either in the central grain or in the two boundary grains. Nevertheless, in the vicinity of the GBs, the plastic strain evolution is captured significantly better, compared to the previous results with Voce hardening. It is notable that for both GP simulations with GB hardening, identical parameter values are used (see Table 4.5). Although the evolution of the plastic strain values near the GBs is captured comparably well, the gradients of the plastic strain near the GBs are underestimated. In some of the grains, however, the GP results overestimate the values from the DDD results (see Fig. 4.3b and Fig. 4.3d).

From a discrete dislocation dynamics point of view, the gradients are –similarly to the plastic strain values within the grains– dependent on the interactions of the dislocations. Therefore, what appears to be missing in the GP model is an appropriate mechanism to consider the elastic interactions of the dislocations across the GBs. These influence the resulting plastic strain profiles (Stricker et al., 2016) and lead to the different distributions in the three grains. It can be shown, using DDD simulations, that the stress-interaction of dislocations is the main contributor to the observed inhomogeneities of the plastic strain profiles (Stricker, 2016). In the current GP model, however, the spatial distributions of plastic strain are mainly influenced by the chosen form of the defect energy and by the GB energy.

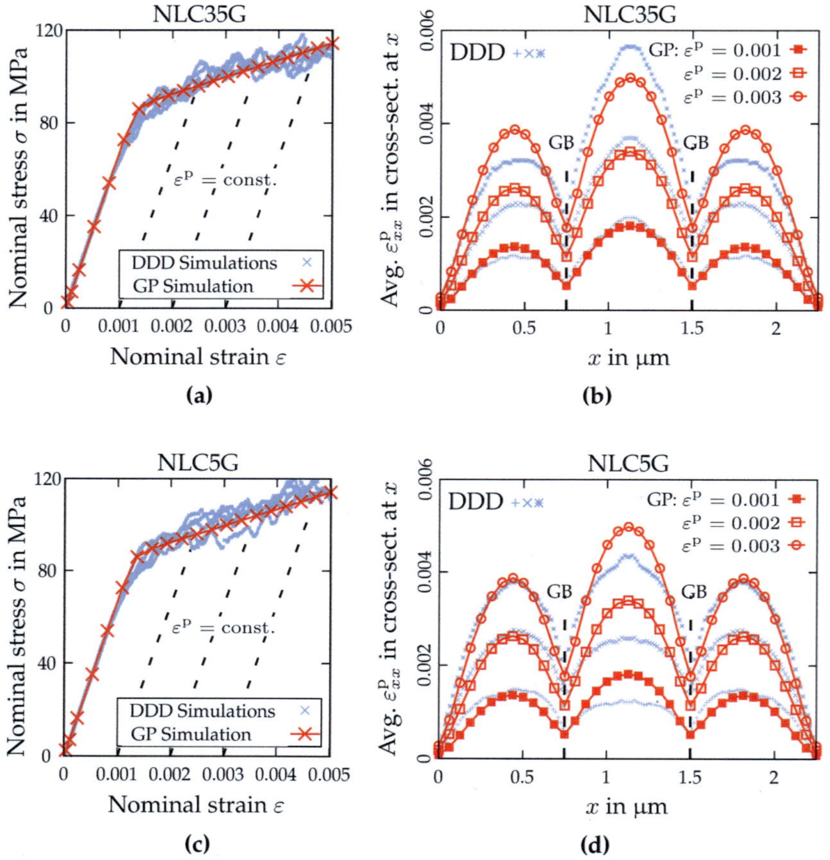

Figure 4.3: (a) Stress-strain curves of DDD and GP simulations for 35° misorientation. (b) Distribution of cross-section averaged plastic strain along loading direction of DDD and GP simulations for 35° misorientation, obtained at fixed overall plastic strain values (a). (c,d) GP and DDD simulation results for 5° misorientation. GP simulations obtained with quadratic defect energy. Central grain rotated by misorientation angle around loading axis. Figures reprinted from Bayerschen et al. (2015).

Therefore, in the following section, an additional case is investigated which allows to isolate the resulting DDD strain profiles from the interactions of dislocations across GBs. Thereby, in the GP model, the influence of the defect energy can be investigated isolated from the influence of the GB energy form. Thus, it can be evaluated if the chosen defect energy of the GP model is sufficient to model the gradients and plastic strain distributions of the DDD results reasonably.

4.3.4 Special case: elastic boundary grains using solely gradient hardening

The material parameters of the two boundary grains of the considered tricrystal are adjusted to represent a purely elastic behavior. Only the central grain is elastic-plastic and is assigned the $\langle 100 \rangle$-orientation with respect to the x-axis (cf. Table 4.6). The GB contributions and the Voce-hardening relation are neglected, and lateral contraction of the boundary planes is not restricted (see Fig. 3.1b for the respective BCs).

Table 4.6: Setup and model parameters of GP simulation for comparison to DDD results. The abbreviation LC indicates that lateral contraction is allowed for on the boundary planes at $x = 0$ and $x = x_{\max}$. In the special case LC0E, the two boundary grains are purely elastic. Only gradient (grad.) hardening is considered. Data reprinted from Bayerschen et al. (2015).

Name	Angle φ	Hardening	$\Xi^{C}_{0,\Gamma}$ (N/m)	$\Xi^{C}_{0,\partial\mathcal{B}}$ (N/m)	τ^{C}_{0} (MPa)	τ^{C}_{∞} (MPa)	Θ (MPa)
LC0E	0°	Grad.	-	-	44.0	-	-

Compared to the previously investigated tricrystal-settings, the overall elastic response of the DDD simulations is slightly harder. This is due the confinement of the plastic activity to the central grain. Consequently, for this special case, the Young's modulus, used in the GP model, has to be adjusted (to a value of 69.4 GPa). The resulting initial defect energy density is $W_0 = 0.5G = 12.88$ GPa. In all other cases, however, the microplasticity-effect (a seemingly smaller Young's modulus) is more pronounced because a small bow-out of a favorably oriented Frank-Read source at the beginning of the simulations is more likely to occur. When GB effects and hardening of the bulk material are neglected in the GP model, the defect energy exclusively controls the overall rate of hardening. The associated normalization constant is fitted to a value of $g_0 = 17.51/\mu\text{m}$. In Fig. 4.4a, the stress-strain results are contrasted, and in Fig. 4.4b a comparison of the strain profiles is shown. The used quadratic form of the defect energy in the GP model leads to an overestimation of plastic strain in the center of the grain. Near the GBs, the gradients are underestimated by the GP results.

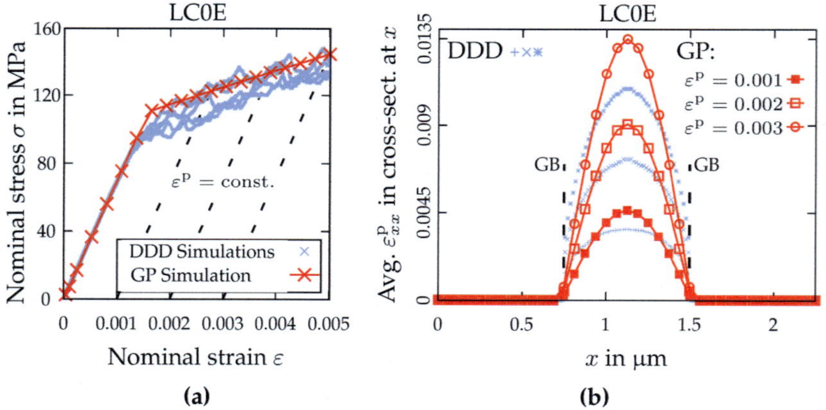

Figure 4.4: (a) Stress-strain curves of DDD and GP simulations for 0° misorientation. Boundary conditions with free lateral contraction. The boundary grains behave elastic (i.e., $\tau_0^C \to \infty$ in the GP model of these). (b) Distribution of cross-section averaged plastic strain along loading direction of DDD and GP simulations, obtained at fixed overall plastic strain values (a). GP simulations obtained with quadratic defect energy. Figures reprinted from Bayerschen et al. (2015).

The DDD profiles, however, show more plateau-like distributions of plastic strain with larger gradients near the GBs. At the same time, the overall shapes of the plastic strain profiles from both simulation approaches differ more pronounced from each other, compared to the other cases above.

It is recalled that in the GP model, the gradients enter the theoretical framework in the free energy via the defect energy contribution, see Eq. (3.8). In order to improve the GP results regarding their prediction of the plastic strain gradients near the GBs and the distributions of plastic strain within the grains, a more general type of defect energy is discussed and tested in the following chapter.

Chapter 5

Extension of the gradient crystal plasticity theory by a power-law defect energy

5.1 Motivation

Gradient plasticity modeling of size-effect related phenomena is influenced by the choice of the mathematical form of the defect energy. It is known that depending on, e.g., the exponent of the defect energy (Bardella, 2010), for example, the magnitude of the modeled size effects can be varied. On the path towards refining the single-crystal gradient plasticity model by Wulfinghoff et al. (2013) and by Bayerschen et al. (2015), a power-law defect energy is realized in this chapter. This (in general) non-linear defect energy is formulated using the gradient $\nabla\zeta$ of the micromorphic counterpart ζ to the accumulated plastic slip γ_{ac}. The approach allows to investigate different exponents of the defect energy in numerical simulation studies. Therefore, also different plastic strain distributions can be modeled which deviate from the strictly parabolic distributions obtained with a quadratic defect energy (see, for example, Section 4.3.4). The content of the following chapter is largely taken from Bayerschen and Böhlke (2016) and Bayerschen et al. (2016c).

5.2 Mathematical model

5.2.1 Preliminaries

In the following, the basic gradient-plasticity framework from Chapter 3 is considered but Voce-hardening is neglected. Therefore, the field equations of the theory used for the bulk material in this chapter are identical to the field equations already presented in Table 3.1. The GB conditions and the BCs, however, are different. For brevity of the following derivations and the basic model studies, the GB contributions are neglected, at first (i.e., $W_\Gamma = 0$). Instead, micro-hard boundary conditions and GB conditions are used. Subsequent to the derivations and the basic model studies, however, the GB contributions are considered using the GB-hardening extended formulation from Chapter 4. Additionally, a comparison of GP results obtained with different defect energy exponent choices to DDD results is presented.

5.2.2 Principle of virtual power and field equations

The virtual internal power is assumed to be given by

$$\delta \mathcal{P}_{\text{int}} = \int_{\mathcal{B}} \left(\boldsymbol{\sigma} \cdot \delta \dot{\boldsymbol{\varepsilon}} + \pi \delta \dot{\zeta} + \boldsymbol{\xi} \cdot \nabla \delta \dot{\zeta} \right) \, \mathrm{d}v, \tag{5.1}$$

with the virtual strain rates $\delta\dot{\boldsymbol{\varepsilon}}(\delta\dot{\boldsymbol{u}})$, and the virtual rate of the micromorphic variable $\delta\dot{\zeta}$. It is assumed that ζ is a continuous quantity. Consequently, possible jumps of ζ, e.g., at grain boundaries in the body $\mathcal{B}$, are neglected. This assumption is not expected to be transferable to plastic slips of individual slip systems, in general. In the context of an overall description of plastic slip, however, it is noted that the calculations of an effective plastic strain in the experimental work of Abuzaid et al. (2012) resulted in continuous distributions of this overall quantity across many grain boundaries. Thus, it appears to be reasonable to consider ζ as a continuous quantity, for a first approach. Furthermore, the virtual power of the external forces is assumed to have the two contributions

$$\delta\mathcal{P}_{\text{ext}} = \int\limits_{\partial\mathcal{B}_{\text{t}}} \bar{\boldsymbol{t}} \cdot \delta\dot{\boldsymbol{u}}\,\text{da} + \int\limits_{\partial\mathcal{B}_{\Xi}} \bar{\Xi}\,\delta\dot{\zeta}\,\text{d}a, \tag{5.2}$$

where power is expended at the external boundaries $\partial\mathcal{B}_{\text{t}} \cup \partial\mathcal{B}_{\Xi} = \partial\mathcal{B}$ by the tractions $\bar{\boldsymbol{t}}$, and by the microtractions $\bar{\Xi}$, respectively. In Eq. (5.2), it has been exploited, already, that the virtual rates $\{\delta\dot{\boldsymbol{u}}, \delta\dot{\zeta}\}$ vanish for given $\{\boldsymbol{u}, \zeta\}$ at the Dirichlet boundaries $\partial\mathcal{B}_{\text{u}}$. By using the principle of virtual power, i.e., by letting the virtual power of the internal forces be equal to the virtual power of the external forces, $\delta\mathcal{P}_{\text{int}} = \delta\mathcal{P}_{\text{ext}}$, and applying Gauss' theorem, the field equations and the Neumann boundary conditions, listed in Table 5.1, can be derived (cf. Appendix H).

Table 5.1: Field equations and boundary conditions with gradient extension neglecting grain boundary contributions. Table reprinted from Bayerschen and Böhlke (2016) with permission from Springer.

Linear momentum balance	$\boldsymbol{0}$	$= \text{div}\,(\boldsymbol{\sigma})$	$\forall \boldsymbol{x} \in \mathcal{B}$
Microforce balance	π	$= \text{div}\,(\boldsymbol{\xi})$	$\forall \boldsymbol{x} \in \mathcal{B}$
Neumann BCs for: Cauchy stress	$\boldsymbol{\sigma}\boldsymbol{n}$	$= \bar{\boldsymbol{t}}$	on $\partial\mathcal{B}_t$
Gradient stress	$\boldsymbol{\xi} \cdot \boldsymbol{n}$	$= \bar{\Xi}$	on $\partial\mathcal{B}_{\Xi}$

5.2.3 Constitutive equations with a power-law defect energy

The free energy of the bulk is assumed to be given by

$$W(\boldsymbol{\varepsilon}, \hat{\lambda}, \zeta, \nabla\zeta) = W_{\text{e}}(\boldsymbol{\varepsilon}, \boldsymbol{\varepsilon}^{\text{p}}(\hat{\lambda})) + W_{\text{g}}(\nabla\zeta) + W_{\chi}(\zeta - \gamma_{\text{ac}}(\hat{\lambda})), \tag{5.3}$$

where the defect contribution W_{g} to the free energy, Eq. (3.9), is substituted by a defect energy of power-law form,

$$W_{\mathrm{g}}(\nabla\zeta) = W_0 \left(\frac{|\nabla\zeta|}{g_0}\right)^m , \tag{5.4}$$

with the defect-energy exponent m. In the present work, the range $1 \leq m \leq 2$ is considered. It is referred to Forest (2016) for a recent overview of different types of defect-energy formulations with a focus on micromorphic theories.

The total dissipation, neglecting GB contributions, reads

$$D_{\mathrm{tot}} = \int\limits_{\mathcal{B}} \mathcal{D}\,\mathrm{d}v = \mathcal{P}_{\mathrm{ext}} - \int\limits_{\mathcal{B}} \dot{W}\,\mathrm{d}v \geq 0, \tag{5.5}$$

if thermal effects are disregarded, too. By setting the external power to be equal to the internal power,

$$\mathcal{P}_{\mathrm{ext}} = \mathcal{P}_{\mathrm{int}} = \int_{\mathcal{B}} \left(\boldsymbol{\sigma} \cdot \dot{\boldsymbol{\varepsilon}} + \pi\dot{\zeta} + \boldsymbol{\xi} \cdot \nabla\dot{\zeta}\right) \mathrm{d}v, \tag{5.6}$$

and substituting Eq. (5.3) in Eq. (5.5), the dissipation $\mathcal{D}$ can be reformulated by the form

$$\begin{aligned}\mathcal{D} = \left(\boldsymbol{\sigma} - \frac{\partial W_{\mathrm{e}}}{\partial \boldsymbol{\varepsilon}}\right) \cdot \dot{\boldsymbol{\varepsilon}} - \frac{\partial W_{\mathrm{e}}}{\partial \boldsymbol{\varepsilon}^{\mathrm{p}}} \cdot \dot{\boldsymbol{\varepsilon}}^{\mathrm{p}} - \frac{\partial W_{\chi}}{\partial \gamma_{\mathrm{ac}}}\dot{\gamma}_{\mathrm{ac}} + \left(\pi - \frac{\partial W_{\chi}}{\partial \zeta}\right)\dot{\zeta} \\ + \left(\boldsymbol{\xi} - \frac{\partial W_{\mathrm{g}}}{\partial \nabla\zeta}\right) \cdot \nabla\dot{\zeta} \geq 0.\end{aligned} \tag{5.7}$$

Using the abbreviation $\check{p} = \partial W_{\chi}/\partial\gamma_{\mathrm{ac}} = -\partial W_{\chi}/\partial\zeta$, assuming the stresses $\boldsymbol{\sigma}$, π, and $\boldsymbol{\xi}$ to be purely energetic, and substituting $\partial W_{\mathrm{e}}/\partial\boldsymbol{\varepsilon}^{\mathrm{p}} = -\boldsymbol{\sigma}$ in Eq. (5.7), leads to the reduced dissipation inequality in the form

$$\mathcal{D} = \boldsymbol{\sigma} \cdot \dot{\boldsymbol{\varepsilon}}^{\mathrm{p}} - \check{p}\,\dot{\gamma}_{\mathrm{ac}} \geq 0. \tag{5.8}$$

Substituting both the plastic strain tensor from Eq. (2.5) and the accumulated plastic slip, Eq. (2.6), in Eq. (5.8) gives

$$\mathcal{D} = \sum_{\alpha} (\tau_\alpha - \breve{p})\, \dot{\lambda}_\alpha \geq 0, \tag{5.9}$$

with the resolved shear stresses $\tau_\alpha = \boldsymbol{\sigma} \cdot \boldsymbol{M}^s_\alpha$. Requiring equality of the right-hand side of Eq. (5.9) to the right-hand side of Eq. (2.37), the dissipative shear stresses can be expressed by

$$\tau^{\mathrm{d}}_\alpha = \tau_\alpha - \breve{p}. \tag{5.10}$$

Finally, using $\breve{p} = -\pi$ from Eq. (5.7) and the microforce balance ($\pi = \mathrm{div}\,(\boldsymbol{\xi})$, see Table 5.1), gives

$$\tau^{\mathrm{d}}_\alpha = \tau_\alpha + \mathrm{div}\,(\boldsymbol{\xi})\,. \tag{5.11}$$

The flow rule of the bulk material is assumed to be of the same form as in Eq. (4.1), i.e.,

$$\dot{\lambda}_\alpha = \dot{\gamma}_0 \left\langle \frac{\tau^{\mathrm{d}}_\alpha - \tau^{\mathrm{C}}_0}{\tau^{\mathrm{D}}} \right\rangle^p = \dot{\gamma}_0 \left\langle \frac{\tau_\alpha - (\tau^{\mathrm{C}}_0 - \mathrm{div}\,(\boldsymbol{\xi}))}{\tau^{\mathrm{D}}} \right\rangle^p .$$

Since solely the gradient-stress hardening contribution is present in the flow rule, due to the neglect of other bulk-material hardening and GB-hardening contributions, the overall hardening is only influenced by this gradient-hardening term $-\mathrm{div}\,(\boldsymbol{\xi})$.

5.3 Exact solution of a laminate for single slip

In order to investigate the distributions of the micromorphic variable ζ that can be obtained with the power-law defect energy for different choices of the exponent m, an exact solution is briefly presented, at first. Subsequently, results of this solution are discussed. By combining Eqs. (3.13) and (5.4), the gradient stress is given as

$$\boldsymbol{\xi} = \frac{\partial W_{\mathrm{g}}(\nabla\zeta)}{\partial\nabla\zeta} = m\frac{W_0}{g_0}\left(\frac{|\nabla\zeta|}{g_0}\right)^{m-1}\frac{\nabla\zeta}{|\nabla\zeta|}. \tag{5.12}$$

Single slip is considered for a simple shear deformation. Then, in the limit case of identical values of the micromorphic variable and the accumulated plastic slip, $\zeta = \gamma_{\mathrm{ac}} = \gamma$ with $\gamma = \gamma(x)$. Consequently, $\nabla\gamma \sim \boldsymbol{e}_x$, and the x-component of $\boldsymbol{\xi}$ reads

$$\boldsymbol{\xi}\cdot\boldsymbol{e}_x = \xi_x = m\frac{W_0}{g_0}\left(\frac{\left|\frac{\mathrm{d}\gamma}{\mathrm{d}x}\right|}{g_0}\right)^{m-1}. \tag{5.13}$$

The term div ($\boldsymbol{\xi}$) in Eq. (5.11) reduces to

$$\mathrm{div}\,(\boldsymbol{\xi}) = \frac{\mathrm{d}\xi_x}{\mathrm{d}x}, \tag{5.14}$$

which, in combination with Eq. (5.13) gives

$$\frac{\mathrm{d}\xi_x}{\mathrm{d}x} = (m-1)m\frac{W_0}{g_0^2}\left(\frac{\left|\frac{\mathrm{d}\gamma}{\mathrm{d}x}\right|}{g_0}\right)^{m-2}\frac{\mathrm{d}\left|\frac{\mathrm{d}\gamma}{\mathrm{d}x}\right|}{\mathrm{d}x}, \tag{5.15}$$

where the second derivative can be expressed by

$$\frac{\mathrm{d}\left|\frac{\mathrm{d}\gamma}{\mathrm{d}x}\right|}{\mathrm{d}x} = \mathrm{sg}\left(\frac{\mathrm{d}\gamma}{\mathrm{d}x}\right)\frac{\mathrm{d}^2\gamma}{\mathrm{d}x\mathrm{d}x}. \tag{5.16}$$

For the stationary single slip case, $\tau_\alpha^{\mathrm{d}} = \tau^{\mathrm{d}} = \tau_0^{\mathrm{C}}$ and $\tau_\alpha = \tau$. Thus, Eq. (5.11) yields

$$\frac{\mathrm{d}\xi_x}{\mathrm{d}x} = \tau_0^{\mathrm{C}} - \tau = -\Delta\tau, \tag{5.17}$$

where $\Delta\tau$ denotes the difference between the resolved shear stress τ and the critical shear stress τ_0^C. Combining Eqs. (5.15) and (5.17) results in the following form

$$-\Delta\tau = (m-1)m\frac{W_0}{g_0^2}\left(\frac{\left|\frac{\mathrm{d}\gamma}{\mathrm{d}x}\right|}{g_0}\right)^{m-2}\frac{\mathrm{d}\left|\frac{\mathrm{d}\gamma}{\mathrm{d}x}\right|}{\mathrm{d}x} \tag{5.18}$$

with $x_{\min} \leq 0 \leq x_{\max}$ and $x_{\min} = -x_{\max}$.
A solution for $\gamma(x)$ is obtained, for example, by considering $\operatorname{sg}(\mathrm{d}\gamma/\mathrm{d}x) = +1$ for $-x_{\max} \leq x \leq 0$, using the boundary conditions $\mathrm{d}\gamma/\,\mathrm{d}x(x=0) = 0$ as well as $\gamma(x = -x_{\max}) = 0$. The corresponding solution reads

$$\gamma(x) = \frac{W_0(m-1)}{\Delta\tau}\left(\frac{\Delta\tau g_0}{W_0 m}\right)^{\frac{m}{m-1}}\left(-\left|x\right|^{\frac{m}{m-1}} + x_{\max}^{\frac{m}{m-1}}\right). \tag{5.19}$$

Analogously, the solution for $0 \leq x \leq x_{\max}$ can be obtained by considering $\operatorname{sg}(\mathrm{d}\gamma/\mathrm{d}x) = -1$ and the two BCs $\mathrm{d}\gamma/\,\mathrm{d}x(x=0) = 0$ and $\gamma(x = x_{\max}) = 0$, respectively.
In addition to the $\gamma(x)$-distributions, the shear-stress evolution is derived, in dependence of the average plastic slip $\bar{\gamma}$. Therefore, the following integral average is considered

$$\bar{\gamma} = \frac{1}{\Delta x}\int_{x_{\min}}^{0} \gamma(x)\mathrm{d}x \tag{5.20}$$

on the interval $x \in [x_{\min}, 0]$ with $x_{\min} = -x_{\max}$ and $\Delta x = -x_{\min}$. The evaluation of Eq. (5.20), considering an average shear stress difference of $\Delta\tau = \Delta\bar{\tau}$, yields

$$\bar{\gamma} = \frac{W_0(m-1)}{\Delta\bar{\tau}}\left(\frac{\Delta\bar{\tau} g_0}{W_0 m}\right)^{\frac{m}{m-1}}\frac{m}{2m-1}x_{\max}^{\frac{2m-1}{m-1}}. \tag{5.21}$$

Solving Eq. (5.21) for $\Delta\bar{\tau}$ and letting the maximal coordinate $x_{\max} = l$ leads to

$$\Delta\bar{\tau}(\bar{\gamma}) = \bar{\tau} - \tau_0^{\mathrm{C}} = \frac{W_0 m}{l^{2m-1} {g_0}^m} \left(\frac{2m-1}{m-1} \bar{\gamma} \right)^{m-1} . \tag{5.22}$$

In order to compare shear distributions for different defect energy exponents, it is assumed that the same average shear $\bar{\gamma} = 0.01$ is present for the different defect energy exponents m. Then, the corresponding values of the dimensionality constants g_0 for different m can be obtained from Eq. (5.21). This gives the values indicated in Table 5.2.

Table 5.2: Different defect energy exponents m and corresponding normalization constants g_0 with respective internal length scales $1/g_0$, values of g_0 obtained from Eq. (5.21) under the assumption that $\bar{\gamma} = 0.01$ for the chosen values of m. Table reprinted from Bayerschen and Böhlke (2016) with permission from Springer.

m	2.0	1.5	1.1
g_0 in $1/\mu\mathrm{m}$	0.39	0.83	2.07
$1/g_0$ in $\mu\mathrm{m}$	2.56	1.2	0.48

For the initial defect energy density, $W_0 = \bar{\alpha} G$ is used with $\bar{\alpha} = 0.5$ and the shear modulus $G = 25$ GPa. In Fig. 5.1a, the normalized distributions of $\gamma(x)$ are depicted for different defect energy exponents m, assuming a shear stress difference of $\Delta\tau = 5$ MPa. It can be observed that the smaller the defect energy exponent m and the internal length scale $1/g_0$, the larger are the developing gradients of $\gamma(x)$ close to the boundaries. A quadratic defect energy leads to a parabolic distribution of $\gamma(x)$. For choices of m closer to one, however, more plateau-like distributions of $\gamma(x)$ are obtained. The resulting evolution of the average shear stress difference $\Delta\bar{\tau}$ is plotted for the plastic range in Fig. 5.1b. The quadratic defect energy leads to uniform, linear hardening, while choices of m closer to one give highly non-linear, at the onset of plasticity more pronounced, but subsequently less pronounced, hardening.

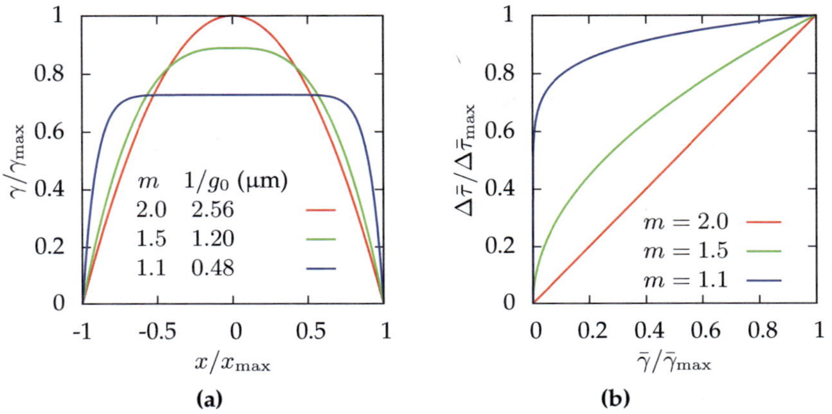

Figure 5.1: (a) Normalized γ-distribution of analytical solution for single-slip simple shear, plotted along the normalized coordinate. Identical $\bar{\gamma}$ for all three values of m. (b) Normalized average shear stress difference, plotted in the plastic range over normalized average plastic slip. Figure reprinted from Bayerschen and Böhlke (2016) with permission from Springer.

If Eq. (5.22) is evaluated for the case $m = 1$,

$$\bar{\tau}(\bar{\gamma}) - \tau_0^{\mathrm{C}} = \frac{W_0}{l g_0} \Rightarrow \bar{\tau}(\bar{\gamma}) = \frac{W_0}{l g_0} + \tau_0^{\mathrm{C}}. \tag{5.23}$$

For this case, the influence of the grain size l and of the dimensionality constant g_0 on the average shear stress can be interpreted in an appealing manner. The average shear stress $\bar{\tau}$ scales inversely linear with both, l and g_0. For exponents $m > 1$, however, these dependencies are non-linear, see Eq. (5.22). This explains the increase in yield stress occurring for smaller values of m in conjunction with smaller grain sizes l that can be observed in the numerical results, e.g., in Fig. 5.4a.
In the following, the influence of the defect energy exponent m is separated from the influence of the internal length scale $1/g_0$ for the illustrative example above. A variation of the values of the dimensionality constant g_0 from Table 5.2 by $\pm 5\%$ is performed. In Fig. 5.2, the resulting

normalized average shear distributions (Fig. 5.2a) and the resulting normalized average shear stress difference courses (Fig. 5.2b) are depicted.

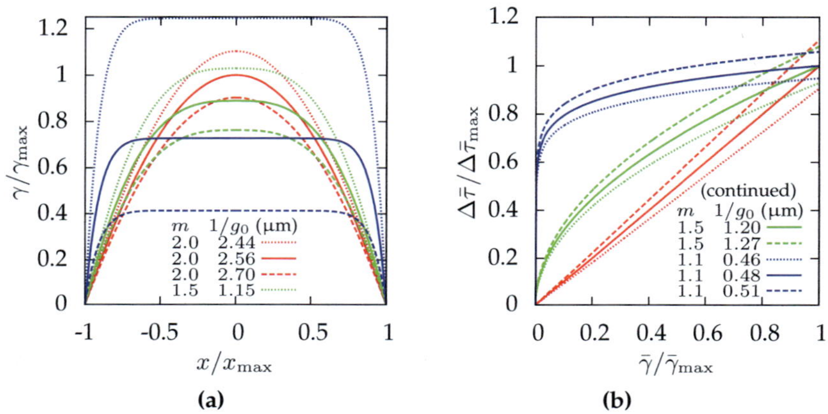

Figure 5.2: (a) Normalized γ-distribution of analytical solution for single slip simple shear, plotted along the normalized coordinate. The normalization constant g_0 is varied as indicated. (b) Normalized average shear stress difference, plotted in the plastic range over normalized average plastic slip. The depicted keys in (a,b) apply to both figures. Figure reprinted from Bayerschen and Böhlke (2016) with permission from Springer.

The courses from Fig. 5.1 are indicated by solid lines, as a reference. From Fig. 5.2a, it becomes evident that the resulting shear is highly non-linear in the variations of m and g_0, respectively. Changes in the defect energy exponent m further amplify variations in the gradients of γ, when changing the internal length scale $1/g_0$. For constant m, an increase in $1/g_0$ leads to more pronounced hardening behavior (see Fig. 5.2b), consistent with less plastic shear (see Fig. 5.2a). Changing the defect exponent m, however, gives rise to substantially different shapes of the respective hardening courses (see Fig. 5.2b).

5.4 Finite element implementation

5.4.1 Linearization of the principle of virtual power

The theory from Section 5.2 is implemented in the in-house FE-code to perform a first numerical investigation with the power-law defect energy. In the following, the key concepts for the implementation of the power-law defect energy approach, Eq. (5.4), and the linearization of the principle of virtual power are discussed, briefly.
The linearization of Eqs. (5.1) and (5.2) in $\delta\dot{\boldsymbol{u}}$ reads

$$\int_{\mathcal{B}} \left(\frac{\partial \boldsymbol{\sigma}}{\partial \boldsymbol{\varepsilon}}[\Delta\boldsymbol{\varepsilon}] + \frac{\partial \boldsymbol{\sigma}}{\partial \zeta}\Delta\zeta \right) \cdot \delta\dot{\boldsymbol{\varepsilon}}\,\mathrm{d}v = -\int_{\mathcal{B}} \boldsymbol{\sigma} \cdot \delta\dot{\boldsymbol{\varepsilon}}\,\mathrm{d}v + \int_{\partial\mathcal{B}_{\mathrm{t}}} \bar{\boldsymbol{t}} \cdot \delta\dot{\boldsymbol{u}}\,\mathrm{d}a \quad \forall \delta\dot{\boldsymbol{u}}. \tag{5.24}$$

Furthermore, the linearization of Eq. (5.1) and Eq. (5.2) in $\delta\dot{\zeta}$, with consideration of $\pi = -\check{p}$, gives

$$\int_{\mathcal{B}} \left(\delta\dot{\zeta}\Big(-\frac{\partial \check{p}}{\partial \zeta}\Big)\Delta\zeta - \delta\dot{\zeta}\frac{\partial \check{p}}{\partial \boldsymbol{\varepsilon}} \cdot \Delta\boldsymbol{\varepsilon} + \frac{\partial \boldsymbol{\xi}}{\partial \nabla\zeta}[\nabla(\Delta\zeta)] \cdot \nabla(\delta\dot{\zeta}) \right) \mathrm{d}v =$$
$$-\int_{\mathcal{B}} \left(-\check{p}\,\delta\dot{\zeta} + \boldsymbol{\xi} \cdot \nabla(\delta\dot{\zeta}) \right) \mathrm{d}v + \int_{\partial\mathcal{B}_{\Xi}} \bar{\Xi}\,\delta\dot{\zeta}\,\mathrm{d}a. \tag{5.25}$$

The numerical integration of Eq. (5.24) and Eq. (5.25) is outlined in Wulfinghoff et al. (2013) in a more general formulation, including GB contributions (see also Section 3.4). Clearly, the FE implementation of Eq. (5.24) and Eq. (5.25) requires to compute the gradient stress $\boldsymbol{\xi} = \partial W_{\mathrm{g}}(\nabla\zeta)/\partial\nabla\zeta$ and its derivative $\partial\boldsymbol{\xi}/\partial\nabla\zeta$. Using the power-law defect energy formulation from Eq. (5.4), the gradient-stress can be rewritten by

$$\boldsymbol{\xi} = \frac{\partial W_{\mathrm{g}}(\nabla\zeta)}{\partial\nabla\zeta} = m\frac{W_0}{g_0^m}|\nabla\zeta|^{m-2}\nabla\zeta. \tag{5.26}$$

The derivative of the gradient stress with respect to the gradient of the micromorphic variable reads

$$\frac{\partial \boldsymbol{\xi}}{\partial \nabla \zeta} = m \frac{W_0}{g_0^m} |\nabla \zeta|^{m-2} \boldsymbol{I} + m(m-2) \frac{W_0}{g_0^m} |\nabla \zeta|^{m-4} \nabla \zeta \otimes \nabla \zeta. \tag{5.27}$$

From Eq. (5.26) and Eq. (5.27) it is apparent that for the choice of an arbitrary power exponent m, the gradient stress in this formulation is neither necessarily defined nor differentiable for the case of vanishing gradients $\nabla\zeta$ (i.e., for $|\nabla\zeta| \to 0$). However, for the special case of a quadratic defect energy ($m = 2$), the gradient stress reduces to the form

$$\boldsymbol{\xi} = 2 \frac{W_0}{g_0^2} \nabla \zeta, \tag{5.28}$$

and its derivative with respect to $\nabla\zeta$ is given by

$$\frac{\partial \boldsymbol{\xi}}{\partial \nabla \zeta} = 2 \frac{W_0}{g_0^2} \boldsymbol{I}. \tag{5.29}$$

Consequently, for this special case, the gradient stress is defined and differentiable, even for vanishing gradients $\nabla\zeta$, making it a numerically convenient choice. It is remarked that a quadratic defect energy leads to a gradient stress $\boldsymbol{\xi}$ that is linear in the gradient $\nabla\zeta$. Choosing $m = 1$, however, gives a constant gradient stress $\boldsymbol{\xi}$ (see also Eq. (5.26)).
In order to use the power-law defect energy with exponents different from $m = 2$, a numerical regularization of the power-law defect energy, Eq. (5.4), for vanishing gradients $\nabla\zeta$, is outlined in the following.

5.4.2 Regularization of the power-law defect energy approach

A standard numerical regularization approach (e.g., Trémolières et al., 1981) is used, here, by adding a small constant to the absolute value

of the gradient $\nabla\zeta$. Regularized defect energy associated quantities are denoted by a $\tilde{(\cdot)}$, in the following.
The regularized form of the defect energy is introduced as

$$\tilde{W}_{\mathrm{g}}(\nabla\zeta) = W_0 \left(\frac{\left(|\nabla\zeta|^2 + \epsilon^2\right)^{\frac{1}{2}}}{g_0} \right)^m = \frac{W_0}{g_0^m} \left(|\nabla\zeta|^2 + \epsilon^2\right)^{\frac{m}{2}}, \quad (5.30)$$

and the numerical parameter ϵ is chosen sufficiently small but non-zero (see Appendix F.2 for a parameter study). This choice leads to finite values of the derivatives of Eq. (5.30), also for vanishing gradients $\nabla\zeta$. The gradient stress for the employed regularization is given by

$$\tilde{\boldsymbol{\xi}} = \frac{\partial \tilde{W}_{\mathrm{g}}(\nabla\zeta)}{\partial \nabla\zeta} = m \frac{W_0}{g_0^m} \left(|\nabla\zeta|^2 + \epsilon^2\right)^{\frac{m}{2}-1} \nabla\zeta, \quad (5.31)$$

and its derivative with respect to the gradient reads

$$\frac{\partial \tilde{\boldsymbol{\xi}}}{\partial \nabla\zeta} = m \frac{W_0}{g_0^m} \left(\left(|\nabla\zeta|^2 + \epsilon^2\right)^{\frac{m}{2}-1} \boldsymbol{I} + (m-2)\left(|\nabla\zeta|^2 + \epsilon^2\right)^{\frac{m}{2}-2} \nabla\zeta \otimes \nabla\zeta \right). \quad (5.32)$$

For the special case $m = 2$, this regularized defect energy form recovers the non-regularized formulation, i.e., $\tilde{\boldsymbol{\xi}} = 2W_0\nabla\zeta/g_0^2$, and $\partial\tilde{\boldsymbol{\xi}}/\partial\nabla\zeta = 2W_0\boldsymbol{I}/g_0^2$, respectively. If the regularization parameter ϵ is set to zero, the original power-law formulation, Eq. (5.4), of the defect energy is recovered, i.e., $\tilde{\boldsymbol{\xi}} = \boldsymbol{\xi}$, and $\partial\tilde{\boldsymbol{\xi}}/\partial\nabla\zeta = \partial\boldsymbol{\xi}/\partial\nabla\zeta$, respectively.

5.4.3 Numerical time integration and algorithmic tangent

The numerical time integration requires the determination of the stresses in Eqs. (5.24–5.25) and the computation of the algorithmic tangent.

These computations follow along the approach discussed in Wulfinghoff et al. (2013). An implicit Euler scheme is applied to Eq. (3.15). For details on this see also Wulfinghoff et al. (2013).

5.5 Numerical results using the defect energy with different exponents m

5.5.1 Tricrystal with elastic boundary grains: m-influence

In order to investigate the model behavior for different exponents m in the defect energy approach, Eq. (5.30), in a fully three-dimensional setting, numerical results are obtained using the parameters from Table 5.3.

Table 5.3: Model parameters for tensile test simulations of aluminum-like tricrystals. Table reprinted from Bayerschen and Böhlke (2016) with permission from Springer.

G	ν	W_0	$\bar{\alpha}$	p
25 GPa	0.3	$\bar{\alpha}G$	0.5	20
$\dot{\gamma}_0$	τ_0^{C}	τ^{D}	ϵ	H_χ
10^{-3} 1/s	33.5 MPa	1 MPa	10^{-6}	10^8 MPa

In a first investigation, three choices of the exponent m are considered. For each choice, the normalization constant g_0 is determined by simulations such that the final nominal-stress value at the final nominal strain is identical for all m. Consequently, for all considered exponents, the plasticity carried by the simulation volume is identical after the final time step. The chosen values of m and the corresponding normalization constants are shown in Table 5.4. Using these, the behavior of the model can be compared for the cases of a quadratic defect energy, a defect energy that is closer to being linear, and a defect energy with an exponent that is in the range between these two cases. It is noted that using a defect energy with an exponent of exactly $m = 1.0$ would require further regularization due to an initial jump in the gradient stress at the beginning of plastic deformation (see, e.g., Ohno and Okumura, 2007).

Table 5.4: Used defect energy exponents m and corresponding normalization constants g_0 with respective internal length scales $1/g_0$. Table reprinted from Bayerschen and Böhlke (2016) with permission from Springer.

m	2.0	1.5	1.1
g_0 in $1/\mu m$	8.25	43.5	450.21
$1/g_0$ in μm	1.2×10^{-1}	2.3×10^{-2}	2.2×10^{-3}

Instead, a value of m is investigated that is sufficiently higher to not necessitate additional regularization. The stress-strain curves that are obtained for all three exponents are depicted in Fig. 5.3a. While the quadratic defect energy leads to a linear hardening behavior, choices of lower exponents m, however, reduce the slope of the (in general non-linear) hardening observed. The field distributions of the micromorphic variable ζ along the central line segment in the elastic-plastic grain are depicted for the three choices of m in Fig. 5.3b. It can be seen, there, that the common choice of $m = 2$ for the exponent yields a parabolic profile for ζ in the central grain. A choice closer to $m = 1$, however, leads to a more plateau-like ζ-distribution in the central grain. The gradients close to the grain boundary rise with decreasing m. For the quadratic defect energy, a parabolic distribution of ζ develops. In contrast, the smaller choices of m lead to the initiation of a region with a constant value of ζ where, in large parts, ζ increases uniformly throughout the evolution. The lower the choice of m, the more pronounced is this region.

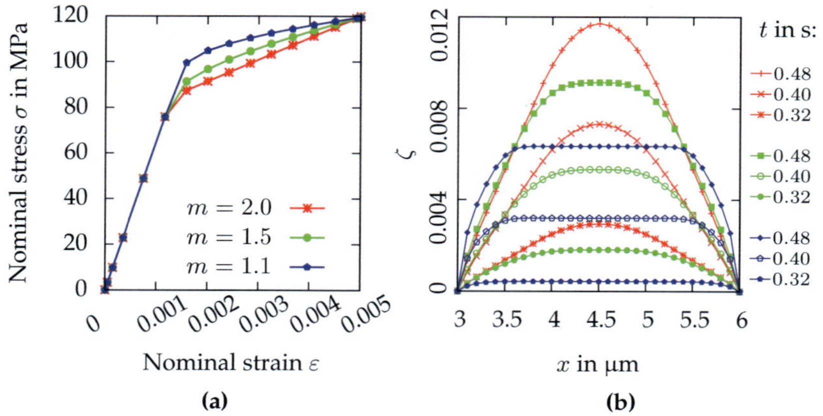

Figure 5.3: (a) Stress-strain curves for tricrystal with elastic boundary grains and different defect energy exponents m. (b) ζ-distributions in the central grain along the x-axis for three time steps. Figure reprinted from Bayerschen and Böhlke (2016) with permission from Springer.

Due to the steeper gradients developing for these values of m, the maximum value that ζ achieves is significantly less than for $m = 2$. The same "amount" of plasticity is carried by the central grain, however, at the end of the last time step of each case.

5.5.2 Tricrystal with elastic boundary grains: size effects

In addition, the size effects occurring for the different defect energy exponents are investigated with the aforementioned tricrystal, considering the same model parameters as in the previous example. The length l of the cubic grains, however, is varied. In Fig. 5.4a, the different size effect behavior is shown for three grain sizes l. Smaller grains respond stiffer for all three values of m, and the magnitude of the size effects is clearly influenced by the choice of the defect energy exponent and the corresponding internal length scales, which are also indicated in Fig. 5.4a. It is remarked that the model behavior is influenced both by the choice of the defect energy exponent and the internal length scale. An increase in the overall yield strength becomes more pronounced for smaller m-values (see also Section 5.3). For constant defect energy exponent and length scale, the tricrystal responds stiffer with decreasing grain size l. A larger magnitude of the size effect is observed for the quadratic choice, compared to the smaller choices of the exponent. The smaller choices of m, however, lead to a less pronounced size effect.

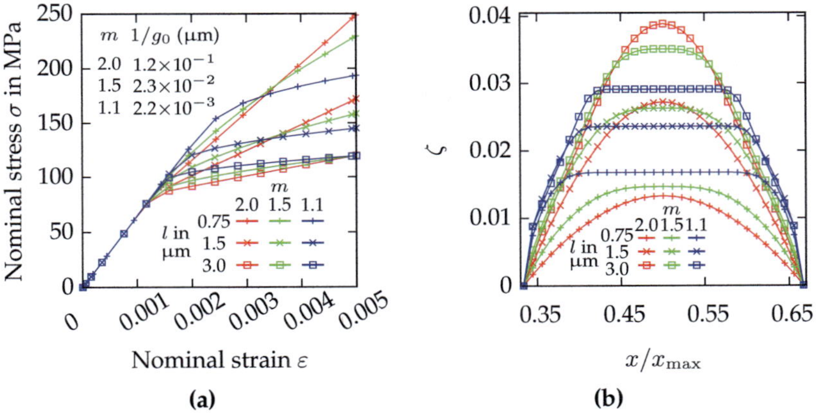

Figure 5.4: (a) Stress-strain curves for tricrystal with elastic boundary grains for different grain sizes l and defect energy exponents m. (b) ζ-distributions in the central grain along the x-axis plotted over normalized coordinate $x/x_{\max}$ at the final time step for different grain sizes l and defect energy exponents m. Figure reprinted from Bayerschen and Böhlke (2016) with permission from Springer.

In addition, the corresponding distributions of the micromorphic variable ζ are plotted at each final time step in Fig. 5.4b. It can be observed, there, that the size of the grains affects the distribution of ζ, significantly. Larger grains tend to show smaller plateau-like distributions and, in turn, larger values of ζ in the center of the grain. The classic size effect of "smaller responding stronger" can be observed in both Fig. 5.4a and

Fig. 5.4b. In detail this means that the smaller the grain size, the smaller is the "amount" of plasticity carried by the central grain, see Fig. 5.4b.

5.5.3 Tricrystal with elastic-plastic grains

As an additional numerical example, a crystal composed of three cubic grains that all behave elastic-plastic is investigated. This shall demonstrate the capability of the employed defect energy approach to be used in elastic-plastic three-dimensional multicrystalline simulations. The material parameters are chosen to be the same as in the previous numerical examples. However, the crystal orientation of the central grain is altered by a rotation of 6° around the z-axis (see the coordinate-system in Fig. 5.5a). The resulting field distributions of ζ are shown in Fig. 5.5a for all three choices of m. Clearly, the width of the gradient-affected zone close to the grain boundaries at $x = 3$ µm, and at $x = 6$ µm, respectively, is influenced by the defect energy exponent m.

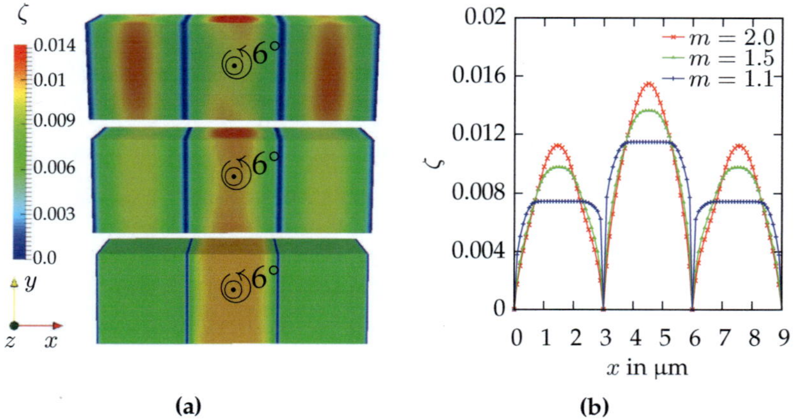

Figure 5.5: (a) Field distributions of ζ for tricrystal with elastic-plastic grains, (b) ζ-distributions along a line segment on the central x-axis for different defect energy exponents m. Figures reprinted from Bayerschen and Böhlke (2016) with permission from Springer.

Lower choices of m lead to steeper gradients of ζ that are distributed over smaller regions. This can also be seen in Fig. 5.5b, where it can be observed, as well, that the differing orientations of the grains lead to different gradients of ζ. For $m = 1.1$, this becomes especially apparent with a very sharp transition in the ζ-distribution, from the plateau in the middle of each boundary grain towards the grain boundaries.

5.6 Comparison of gradient plasticity model results to discrete dislocation dynamics results

5.6.1 Tricrystal with elastic boundary grains

In this section, the GP models of Bayerschen et al. (2015) and Bayerschen and Böhlke (2016) are combined. Specifically, GB contributions are accounted for and GB hardening is considered. Therefore, the field equations are given by the formulation in Table 3.1. The GB yield strength is given by the form Eq. (4.3). For the defect energy, the power-law approach, presented in Section 5.2, is used including the discussed numerical regularization.
At first, the tricrystal with elastic boundary grains is considered in order to compare GP results, obtained with different defect energy exponents m, to the DDD results from Chapter 4. It is recalled that the choice of both the defect energy exponent m and the normalization constant g_0 affect the overall stress-strain response and the plastic strain profiles (see Section 5.3). Therefore, for each choice of the exponent m, the corresponding normalization constant g_0 is adjusted such that the nominal stress at the end of the simulation is approximately the same as the nominal-stress value obtained for $m = 2$ (see Fig. 5.6a).

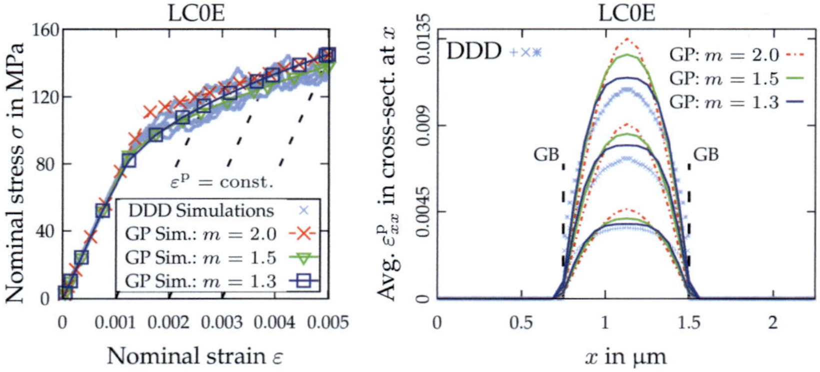

(a) **(b)**

Figure 5.6: (a) Stress-strain curves of DDD and GP simulations for 0° misorientation. Boundary conditions with free lateral contraction. The boundary grains behave elastic (i.e., $\tau_0^C \to \infty$ in the GP model of these). (b) Distribution of cross-section averaged plastic strain along loading direction of DDD and GP simulations, obtained at the fixed overall plastic strain values from (a). GP simulations obtained with different defect energy exponents m. DDD data and GP data for $m = 2.0$ from Bayerschen et al. (2015), GP Data for $m = 1.3$ and basic figures reprinted from Bayerschen et al. (2016c).

Thereby, the set of parameters in Table 5.5 is obtained. The used elastic constants are identical to the elastic constants from Chapter 4. By using defect-energy exponents closer to $m = 1$ a better prediction of the plastic strains and the gradients is achieved for the tricrystal with elastic boundary grains (see Fig. 5.6b). With increasing plastic deformation, however, the deviations between the GP and the DDD results increase.

Table 5.5: Parameters for gradient plasticity simulations of tricrystals with elastic boundary grains (LC0E) and different defect-energy exponents m. Data for $m = 1.3$ from Bayerschen et al. (2016c).

m	$W_0 = G/2$ (GPa)	g_0 (1/µm)	τ_0^C (MPa)
1.5	12.88	133.5	32.5
1.3	12.88	200	24.5

5.6.2 Tricrystal with elastic-plastic grains

For the case of a fully elastic-plastic tricrystal, the stress-strain responses are contrasted for different defect energy exponents and different misorientations of the central grain in Fig. 5.7a and Fig. 5.7c. The used parameter values of the GP model are listed in Table 5.6.

Table 5.6: Parameters for gradient plasticity simulations of tricrystals with elastic-plastic grains (NLC35G/NLC5G) and different defect-energy exponents m. Data for $m = 1.3$ from Bayerschen et al. (2016c).

m	$W_0 = G/2$ (GPa)	g_0 (1/µm)	τ_0^{C} (MPa)	$\Xi_{0,\mathrm{GB}}^{\mathrm{C}}$ (N/m)	$\Xi_{0,\partial\mathcal{B}}^{\mathrm{C}}$ (N/m)	K_{H} (N/m)
1.5	12.06	93.5	27.5	4	25	2.0×10^3
1.3	12.06	240	29.5	5.5	25	1.8×10^3

A better prediction of the gradients of the plastic strain is obtained in the central grain for both misorientation cases using a defect energy exponent of $m = 1.3$ (see Fig. 5.7b and Fig. 5.7d). In the two adjacent grains, however, it is observable that the gradients are underestimated, nevertheless (see also Fig. 5.7b and Fig. 5.7d). This behavior can be attributed to the simplification in the GP model, considering only one gradient stress $\boldsymbol{\xi}$. The jump of the gradient stress is balanced at the GB against the GB yield strength Ξ_0^{C}, thereby limiting the gradients that can develop in the GB vicinity. This model restriction could be relaxed by consideration of all gradients $\nabla\gamma_\alpha$, and the associated higher-order (back)stresses. In that case, the jumps of the gradient-stresses would be balanced against GB microtractions for the individual slip systems. As a consequence, the resulting overall jump of the plastic strain at the GBs would be significantly less restricted. In addition, the misorientation-dependent elastic interactions of dislocations across the GBs (Stricker et al., 2016) are not considered in the GP model.

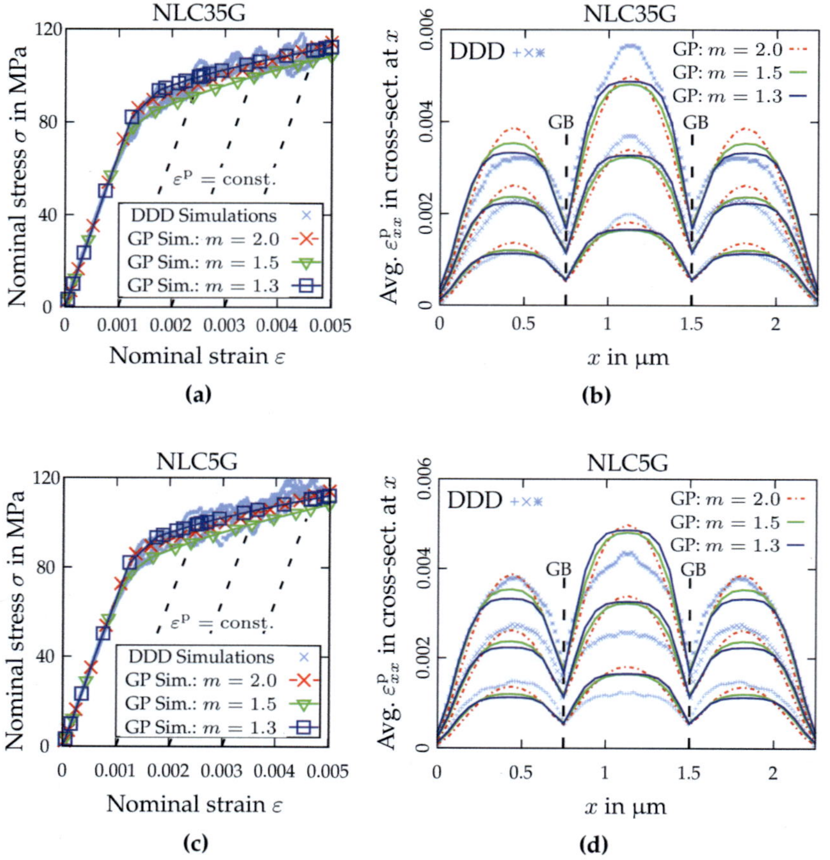

Figure 5.7: (a) Stress-strain curves of DDD and GP simulations for 35° misorientation. (b) Distribution of cross-section averaged plastic strain along loading direction of DDD and GP simulations for 35° misorientation, obtained at the fixed overall plastic strain values from (a). (c,d) GP and DDD simulation results for 5° misorientation. GP simulations obtained with different defect energy exponents m. Central grain rotated by misorientation angle around loading axis. DDD data and GP data for $m = 2.0$ from Bayerschen et al. (2015), GP Data for $m = 1.3$ and basic figures rerprinted from Bayerschen et al. (2016c).

Thus, the strain profile results do not reflect the influence by such interactions (compare Fig. 5.7b to Fig. 5.7d), indicating that the GP model could be further enhanced by an additional mechanism to account for these. In this context, the consideration of dislocation transport (Wulfinghoff and Böhlke, 2015) could also be a remedy in order to model such interactions across GBs appropriately on the continuum scale.

Chapter 6

Towards orientation-dependent modeling of grain boundary slip mechanisms in crystal plasticity models

6.1 Motivation

One of the aims of the present thesis is the consideration of GB plasticity within the used continuum crystal-plasticity framework. A simplified modeling approach has been taken due to the usage of an accumulated plastic slip as additional degree of freedom. This approach does not consider the influence by the orientations of the grains or of the GBs on the transfer of plastic slip across the GBs. As it has been discussed in Section 1.3, however, such mechanisms are an important ingredient that would need to be considered in a refined theory. On the path towards this, some insight is given into this topic, in the following. At first, a brief overview of computational investigations in the context of slip transmission criteria is given. Subsequently, the use of these criteria in crystal-plasticity models is discussed. A model from the literature accounting for the orientation influence on the plastic slip near the GBs (Gurtin, 2008) is connected to some of the slip transmission criteria used in experimental works. The model has recently been implemented

(Gottschalk et al., 2016; McBride et al., 2016) and it is shown, here, how the slip transmission criteria are accounted for within the modeling approach. Additionally, a comparison of the geometrical slip transmission criteria is provided for the single-slip case of two adjacent grains. Thereby, the geometrical criteria from the literature, that have also been used in continuum models, are reviewed. Finally, it is discussed in this chapter why the current model does not appear to be suitable to directly consider the orientation-dependence in the GB approach and suggestions are given for a refinement. The content of Sections 6.2–6.5 is taken from Bayerschen et al. (2016a) and the data for the plots has been prepared by Ramani (2016).

6.2 Computational modeling of orientation-dependent slip transmission at grain boundaries

6.2.1 Computational investigation of slip transmission criteria

The modeling of the dislocation behavior in theories resolving individual dislocations is usually of sophisticated nature regarding the underlying physics. In simulations based on approaches such as atomistics (e.g., Brandl et al., 2007; Bitzek et al., 2009; Bachurin et al., 2010) or discrete dislocation dynamics (e.g., Weygand et al., 2002; Li et al., 2009), for example, mechanisms such as dislocation reactions or the elastic interaction of dislocations across GBs (Stricker et al., 2016) are considered. The slip transmission criteria of combining a geometric transmission factor (GTF) with consideration of the RSS and the RBV (see also Section 1.4) were investigated in atomistic simulations (see Spearot and Sangid (2014), and Bieler et al. (2009) for an overview) and molecular dynamics simulations (Koning et al., 2002). However, it has also been found that the local energetic structure and the local stress state of the GB cannot be neglected, in general, within the context of resolving dislocation interactions near GBs (Spearot and Sangid, 2014). Additional considerations include, e.g., the shear strength of the interface (Demkowicz and Thilly, 2011; Wang et al., 2011; 2012). In the atomistic simulations by Sangid et al. (2012) and in the combined computational / experimental approach by Abuzaid et al. (2012), the importance of the RBV for the slip transmission has been demonstrated. The barrier provided for dislocation motion by two twist and tilt GBs, respectively, was found to be proportional to the RBV magnitude. This was also the case for an investigated twin boundary. The coupled atomistic / discrete dislocation framework Dewald and Curtin (2007a) also confirms the three basic slip transmission criteria (GTF, RSS, RBV) for a tilt GB impinged by edge dislocations. It is proposed there, however, that additional criteria are necessary for the case of GB dislocation nucleation.

For the case of screw dislocations impinging on the same GB type (and other symmetric tilt boundaries), no transmission of dislocations but only nucleation was observed by Dewald and Curtin (2007b). In Dewald and Curtin (2011), dislocations of mixed character are investigated for the same types of boundaries. The set of criteria outlined in Dewald and Curtin (2007b) are extended since the effects of the local GB structure are not accounted for by the classic criteria. They are incorporated, additionally, by quantitative criteria (Dewald and Curtin, 2011).

By such physically detailed simulation approaches such as atomistics, the interaction of dislocations, e.g., with twin boundaries (Ezaz et al., 2011), can be investigated very thoroughly. However, contrary to many continuum models, immense computational costs arise due to the detailed modeling of interactions and the discreteness inherent to the models. This limits using discrete models for larger structures and necessitates the development of, e.g., mesoscale approaches such as gradient-extended crystal plasticity models.

6.2.2 Crystal plasticity models taking into account geometrical slip transmission criteria

The previously described criteria allow for the evaluation of slip transmission prediction by dislocation based crystal plasticity models (Zikry and Kao, 1996). They can also be explicitly incorporated in continuum models to account for the transmission mechanisms. Continuum models, however, lack the discreteness inherent to simulation approaches such as DDD or molecular dynamics. Therefore, the incorporation of dislocation transmission and activation processes near GBs can only be performed in an averaged sense. In the model of Ekh et al. (2011), for example, a functional relationship is proposed for the GB (slip transmission) strength. The strength depends on the minimum angle between the slip directions of slip systems in adjacent grains via

$$\tan\left(\varphi_{\alpha}^{\mathrm{AB}}\right) = \tan\left(\min_{\beta}(\arccos\left(|\boldsymbol{d}_{\alpha}^{\mathrm{A}} \cdot \boldsymbol{d}_{\beta}^{\mathrm{B}}|\right))\right), \tag{6.1}$$

where the slip system convention from Fig. 1.8 is used,here. The higher the minimum angle $\varphi_{\alpha}^{\mathrm{AB}}$, the higher is the GB strength. This criterion, however, does not consider the orientations of the GB normal and of the slip plane normals. The RSSes are accounted for in the flow rule for the slip systems. Thereby, it is ensured that slip systems with large RSSes are activated while others with lower RSSes are not.

In Shi and Zikry (2009; 2011), criterion (1.6) is utilized in combination with a RSS criterion. In case the geometric transmission factor is greater than a critical threshold, and if the ratio of the RSS of an outgoing system with respect to a (with dislocation density evolving) reference shear stress is larger than one, dislocation density can pass the GB in this model and increase the density in the adjacent grain. The thermally activated transmission approach of Ma et al. (2006) assumes that the slip lines of dislocations align with the GB during transmission. There, a criterion is proposed that is based on the minimization of the energy

for a transmission event. This incorporates the RBV in the GB as well as the slip system and the GB orientations. In the employed flow rule, the RSSes are considered and a cutting stress is calculated which models forest dislocations as well as the GB activation energy barrier. Thus, the flow rule connects a minimization of the RBV upon transmission with a maximization of the RSSes, for an activation of plastic slip.
The GB model by Gurtin (2008) has been implemented within a two-dimensional setting in Özdemir and Yalçinkaya (2014) and, recently, also within three dimensions by Gottschalk et al. (2016). In this model, so-called *inter*-action coefficients describe the interactions of slip systems in adjacent grains,

$$\hat{C}^{AB}_{\alpha\beta} = (\boldsymbol{d}^{\mathrm{A}}_{\alpha} \cdot \boldsymbol{d}^{\mathrm{B}}_{\beta})(\boldsymbol{l}^{\mathrm{A}}_{\alpha} \cdot \boldsymbol{l}^{\mathrm{B}}_{\beta}). \tag{6.2}$$

In fact, the *inter*-action coefficients in (6.2) are formally identical to the geometric slip transmission factor (1.6). The model in Gurtin (2008), furthermore, accounts for the RBV criterion and the RSS criterion, as well (see Section 6.3). The superscripts {A,B} distinguish the *inter*-action coefficients from the so-called *intra*-action coefficients. The latter, *intra*-action, coefficients determine the interaction of slip systems within each grain by (6.2), applied to each grain {A,B}, individually, i.e., $\hat{C}^{AA}_{\alpha\beta}$, $\hat{C}^{BB}_{\alpha\beta}$.

6.2.3 Criteria that consider threshold values for the slip system and grain boundary angles

In Ashmawi and Zikry (2002), criterion (1.7) is extended to account for the slip plane normals intersection angle via an additional term

$$\hat{\zeta}_{\alpha\beta} = (\boldsymbol{l}^{\mathrm{A}}_{\alpha} \cdot \boldsymbol{l}^{\mathrm{B}}_{\beta})(\boldsymbol{n}^{\mathrm{A}}_{\alpha} \cdot \boldsymbol{n}^{\mathrm{B}}_{\beta}). \tag{6.3}$$

Critical angles (motivated by Davis et al., 1966; Werner and Prantl, 1990) are used with $\omega_{\mathrm{c}} = 35°$ and $\tilde{\delta}_{\mathrm{c}} = 15°$ (see also Fig. 1.8). The employed

slip transmission factor is purely geometric, but it is combined with the dislocation densities and their evolution on the adjacent sides of GBs. The RSSes are considered in the flow rule. For determining a possible transmission of dislocation density across the GB, the signs of the slip rates are checked, i.e., it is determined whether dislocations in a pile-up move towards the GB or away from it. Thereby, the geometric criterion is connected to the RSSes.
In the work of Mayeur et al. (2015), criterion (1.9) is used to penalize slip transfer on geometrically unfavorable slip system combinations across bimetallic interfaces by increasing the corresponding slip resistances in dependence of the mismatch. The modified slip resistance enters the flow rule and, thus, connects the geometrical factors to the RSSes in the flow rule. The slip transmission criteria used in continuum models are summarized in Table A.2.

6.3 A connection between Gurtin's grain boundary theory and slip transmission criteria used in experiments

The single-crystal plasticity framework of Gurtin (2002) uses the measure of a Burgers tensor field to characterize the Burgers vectors of geometrically necessary dislocations. This measure is defined by

$$\boldsymbol{G} = \mathrm{curl}(\boldsymbol{H}^{\mathrm{p}}) \tag{6.4}$$

for the geometrically linear case, where $\boldsymbol{H}^{\mathrm{p}} = \sum_\alpha \gamma_\alpha \boldsymbol{d}_\alpha \otimes \boldsymbol{n}_\alpha$ is the plastic distortion. Precisely, the defect contribution to the free energy is formulated in dependence of this quantity. As it has been previously mentioned, the need for a defect contribution in the free energy of continuum models results from the coarsening error made by the continuum modeling of the elastic energy (Mesarovic, 2010). By considering the Burgers tensor in the free energy, dependencies of the related higher-order stresses on this measure can be introduced on the individual slip systems, subsequently. These stresses are considered in the flow rule for the slip systems and, thus, influence the mechanical model response.

In Gurtin and Needleman (2005), the theory has been extended by consideration of interfaces such as GBs. The associated boundary conditions are prescribed by the limits of microfree and microhard conditions. Subsequently, the framework by Gurtin (2008) has been developed incorporating the misorientation of adjacent grains and its influence on the slip transfer behavior at the GBs. This is accomplished by considering the Burgers tensor field on the GB. The magnitude $|\boldsymbol{G}|$ is used as a measure of defect in the GB free energy. From the GB energy, internal (energetic) microforces can be derived. These, in turn, are balanced on the GBs with the projections of the vectors of gradient stresses from the adjacent grains. Furthermore, these gradient stresses enter a microforce balance for each slip system α in which the RSSes enter as well.

Although one might expect Gurtin's theory of GBs (Gurtin, 2008) to be connected to criteria of slip system interaction that have been used in experiments, the framework used in the mentioned work is not discussed from this point of view. Therefore, a single slip case is considered, in the work at hand, to show the connections between Gurtin (2008) and the criteria of GTF / RSS / RBV. For convenience, in the following, the single slip systems in grain A and B are labeled by superscripts A and B, respectively. For brevity, the slip plane normals of the two slip systems on adjacent sides of the GB are considered to be coinciding, i.e., $\boldsymbol{n}^{\mathrm{A}} = \boldsymbol{n}^{\mathrm{B}} = \boldsymbol{n}$, and to be perpendicular to the GB normal $\boldsymbol{n}_{\Gamma}$. Thus, the angles $\tilde{\delta} = \omega = 0$, while $\kappa \neq 0$ (see Fig. 1.8). The RBV can be defined as the difference between the Burgers vectors of interacting, i.e., transmitting slip systems (see Lim and Raj, 1985c, and, also, Eq. (1.3)). Its magnitude can be approximated by

$$|\boldsymbol{b}_{\mathrm{r}}| = |\boldsymbol{d}^{\mathrm{A}} - \boldsymbol{d}^{\mathrm{B}}|, \tag{6.5}$$

the magnitude of the difference between the two slip directions $\boldsymbol{d}^{\mathrm{A}}, \boldsymbol{d}^{\mathrm{B}}$ (Abuzaid et al., 2012). Furthermore, the definition of the jump of the plastic distortion $\boldsymbol{H}^{\mathrm{p}}$ across the GB is considered (Gurtin, 2008). For the single slip transmission case at hand, this jump reads

$$[\![\boldsymbol{H}^{\mathrm{p}}]\!] = \gamma^{\mathrm{B}}\boldsymbol{d}^{\mathrm{B}} \otimes \boldsymbol{n} - \gamma^{\mathrm{A}}\boldsymbol{d}^{\mathrm{A}} \otimes \boldsymbol{n} = (\gamma^{\mathrm{B}}\boldsymbol{d}^{\mathrm{B}} - \gamma^{\mathrm{A}}\boldsymbol{d}^{\mathrm{A}}) \otimes \boldsymbol{n}, \tag{6.6}$$

which gives a GB Burgers tensor $\boldsymbol{G}$ (Gurtin, 2008) of

$$\boldsymbol{G} = (\gamma^{\mathrm{B}}\boldsymbol{d}^{\mathrm{B}} - \gamma^{\mathrm{A}}\boldsymbol{d}^{\mathrm{A}}) \otimes (\boldsymbol{n} \times \boldsymbol{n}_{\Gamma}) = (\gamma^{\mathrm{B}}\boldsymbol{d}^{\mathrm{B}} - \gamma^{\mathrm{A}}\boldsymbol{d}^{\mathrm{A}}) \otimes \boldsymbol{l}. \tag{6.7}$$

Assuming, for simplicity, the same plastic slip on both slip systems, i.e., $\gamma^{\mathrm{A}} = \gamma^{\mathrm{B}} = \gamma$, gives a squared magnitude of

$$|\boldsymbol{G}|^2 = \gamma^2(\boldsymbol{d}^{\mathrm{B}} - \boldsymbol{d}^{\mathrm{A}}) \cdot (\boldsymbol{d}^{\mathrm{B}} - \boldsymbol{d}^{\mathrm{A}})\boldsymbol{l} \cdot \boldsymbol{l} = \gamma^2|\boldsymbol{b}_{\mathrm{r}}|^2. \tag{6.8}$$

It is noted that the GB free energy with respect to $|\boldsymbol{G}|$ can, thus, be expressed in dependence of $|\boldsymbol{b}_{\mathrm{r}}|$, the magnitude of the RBV $\boldsymbol{b}_{\mathrm{r}}$, for the special case under consideration. Consequently, Gurtin's GB theory takes into account the residual dislocation content of the GB.
The quantity $|\boldsymbol{G}|^2$ can most generally be expressed by (cf. Gurtin, 2008)

$$|\boldsymbol{G}|^2 = \sum_{\alpha,\beta} \left(C^{\mathrm{AA}}_{\alpha\beta}\gamma^{\mathrm{A}}_{\alpha}\gamma^{\mathrm{A}}_{\beta} + C^{\mathrm{BB}}_{\alpha\beta}\gamma^{\mathrm{B}}_{\alpha}\gamma^{\mathrm{B}}_{\beta} - 2C^{\mathrm{AB}}_{\alpha\beta}\gamma^{\mathrm{A}}_{\alpha}\gamma^{\mathrm{B}}_{\beta} \right), \tag{6.9}$$

which depends on the *intra*-action coefficients $C^{\mathrm{AA}}_{\alpha\beta}, C^{\mathrm{BB}}_{\alpha\beta}$ and on the *inter*-action coefficients $C^{\mathrm{AB}}_{\alpha\beta}$. Following from Eq. (6.9), and from the discussion below Eq. (6.2), it can be concluded that Gurtin's theory of GBs accounts for the geometric slip transmission criterion (1.6).
For the considered case, the *intra*-action coefficients are $C^{\mathrm{AA}} = C^{\mathrm{BB}} = 1$ while all other *intra*-action coefficients vanish. The *inter*-action coefficients vanish, as well, except for $C^{\mathrm{AB}} = C^{\mathrm{BA}} = \boldsymbol{d}^{\mathrm{A}} \cdot \boldsymbol{d}^{\mathrm{B}}$.

Thus, the squared magnitude of the grain boundary Burgers tensor, Eq. (6.9), reads

$$|\boldsymbol{G}|^2 = 2\gamma^2 \left(1 - (\boldsymbol{d}^{\mathrm{A}} \cdot \boldsymbol{d}^{\mathrm{B}})\right). \tag{6.10}$$

Combining (6.10) with (6.8) yields

$$|\boldsymbol{b}_{\mathrm{r}}|^2 = 2 \left(1 - (\boldsymbol{d}^{\mathrm{A}} \cdot \boldsymbol{d}^{\mathrm{B}})\right) \tag{6.11}$$

For the special case of coinciding slip directions, $\boldsymbol{d}^{\mathrm{A}} = \boldsymbol{d}^{\mathrm{B}}$, this gives $|\boldsymbol{b}_{\mathrm{r}}|^2 = 0$, and for the case of perpendicular slip directions, $|\boldsymbol{b}_{\mathrm{r}}|^2 = 2$ is obtained. Thus, the GB RBV magnitude is a function of the mismatch between slip systems in adjacent grains. The GB Burgers tensor magnitude $|\boldsymbol{G}|$ is a function of the mismatch as well, as is the GB free energy of Gurtin (2008) formulated with respect to this quantity.
Concluding, Gurtin's GB theory considers the geometrical slip transmission factor (1.6) in the formulation of the GB free energy (via *interaction coefficients*). They are also incorporated in the formulation of the flow rule by Gurtin (2008). In addition, the RSSes on the outgoing slip systems are considered in the theory (microforce balance / flow rule), as is the RBV left in the GB upon a transmission event (GB free energy).

6.4 Comparison of geometric criteria for single slip

The GTFs from Section 1.4 and from Section 6.2.2 are compared for the single slip case in order to discuss their differences and limitations in more detail with regard to the crystallographic orientation of the grains and of the GB. In the following, it is assumed that plastic slip is occurring on the incoming slip system of grain A and that, subsequently, plastic slip is activated on the outgoing slip system in grain B. Four cases are considered which are depicted in Figs. 6.1a–6.1d.

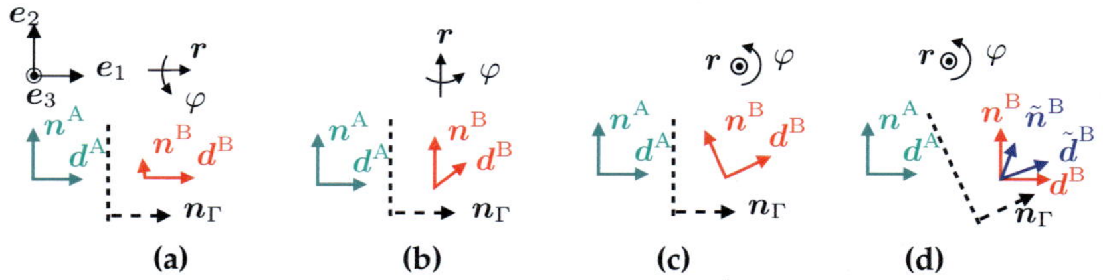

Figure 6.1: (a-d): Rotations employed on a single slip system in grain B and on the grain boundary, respectively, in order to compare the geometric transmission factors. The vectors $\tilde{\boldsymbol{n}}^{\mathrm{B}}$ and $\tilde{\boldsymbol{d}}^{\mathrm{B}}$ are obtained by rotating $\boldsymbol{n}^{\mathrm{B}}$, $\boldsymbol{d}^{\mathrm{B}}$ by $15°$ about the axis $\boldsymbol{a} = \boldsymbol{e}_1 + \boldsymbol{e}_2 + \boldsymbol{e}_3$. Figure reprinted from Bayerschen et al. (2016a) with permission from Springer.

In the cases of Figs. 6.1a–6.1c, the slip system in grain B is rotated by the angle φ about the depicted $\boldsymbol{r}$-axes. In a separate case, the GB is rotated by angle φ about the $\boldsymbol{r}$-axis, depicted in Fig. 6.1d. For this case, at first, the slip system in grain B is left unaltered. Thereby, the influence of the GB inclination can be isolated. Then, the slip system in grain B is pre-rotated by an angle of $15°$ about an arbitrarily chosen axis $\boldsymbol{a} = \boldsymbol{e}_1 + \boldsymbol{e}_2 + \boldsymbol{e}_3$, and the influence of the rotation of the GB is investigated, again. By this approach, a more general case, than the one with ideal alignment of the two slip systems, is considered.
Plots of the geometric transmission factors are depicted in Fig. 6.2. Each column shows results of all geometric factors for the respective cases in Fig. 6.1. Since single slip is investigated, the sums in some of the geometric factors contain only one component. Thus, the weighting with, e.g., Schmid factors is not applicable, there. In the same spirit, for each geometric factor matrix, only the single occurring component is investigated. For a rotation of the slip system B about the slip direction (Fig. 6.1a), the geometric factors (1.4)-(1.7) predict the same behavior. At $\varphi = 90°$, all these factors vanish due to the slip plane normals being perpendicular to each other. Factor (1.11a) also shows essentially the same behavior but its values remain positive due to the multiplication of two (negative) factors.

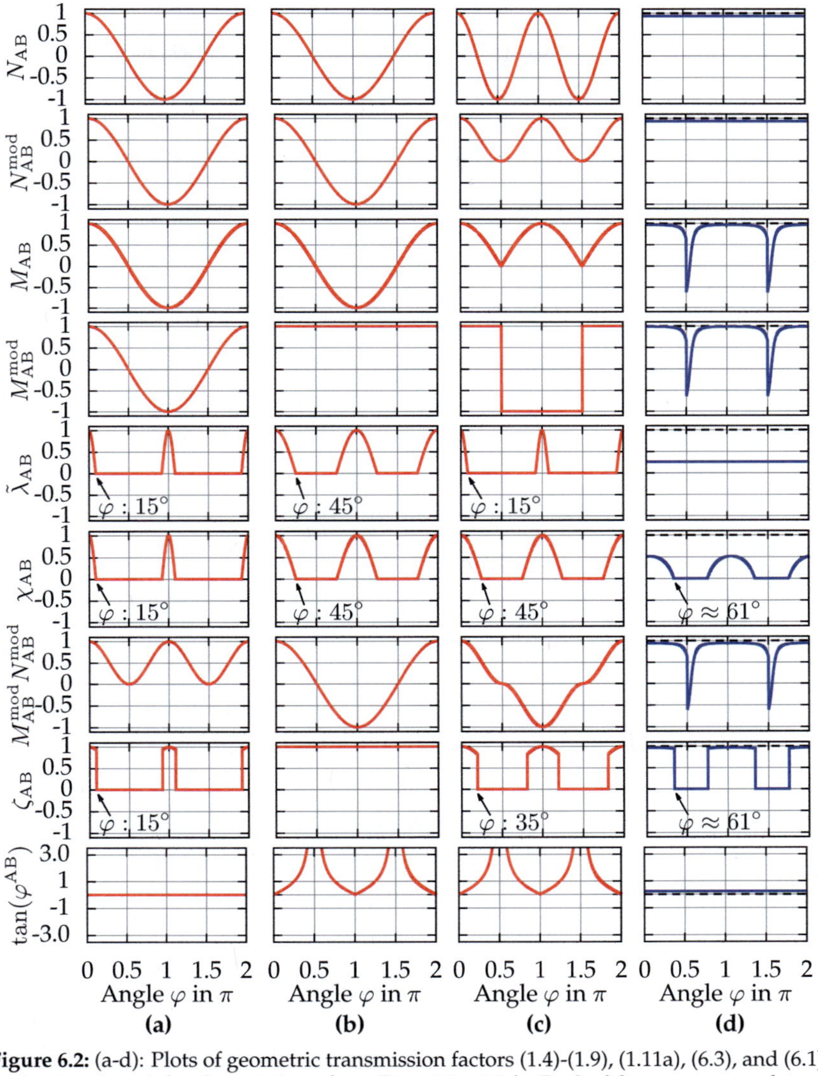

Figure 6.2: (a-d): Plots of geometric transmission factors (1.4)-(1.9), (1.11a), (6.3), and (6.1) (order top-down) for the rotations from Figs. 6.1a–6.1d. Dashed lines correspond to a rotation of the grain boundary for identical slip systems (d), and solid lines to a rotation of the grain boundary for the pre-rotated slip system B, see Fig. 6.1d. Figure reprinted from Bayerschen et al. (2016a) with permission from Springer.

For $\varphi = 180°$, all the preceding factors predict ideal alignment, taking into account opposing slip plane normals. The factors (1.8), (1.9) and (6.3) also vanish at $\varphi = 90°$. The angular region, however, where slip transmission can occur, is far less pronounced than for the previous factors. This is due to the inherent limits, imposed by the critical angles for slip transmission, see, e.g., the plot of factor (1.8) in Fig. 6.2a. In addition, the factors (1.8), (1.9), and (6.3) are different in their transition behavior: while (1.8), (1.9) predict a steep but smooth transition from possible slip transfer to no slip transfer, (6.3) shows a less steep transition at first, then a jump to "no transmission". This is caused by the scaling that is included in (1.8), (1.9) but not in (6.3). The factor (6.1) vanishes for this case. This corresponds to fully possible transmission since (6.1) is used in the resistance of the GB against transmission of plastic slip. However, the applicability of this factor is limited due to the neglect of the slip plane orientations.

Rotating the slip direction by angle φ about the slip plane normal (Fig. 6.1b) gives identical results for factors (1.4)-(1.6) as in the previous case. However, (1.7) is not affected by this rotation since it only considers the lines of intersection with the GB, and not the slip directions. One should, nevertheless, keep in mind that (1.7) is combined with additional considerations regarding the slip directions (see Fig. 1.9). The product (1.11a) of (1.7) and (1.5) is identical to (1.7), as a consequence of (1.5) being constant for the investigated case. The factors (1.8), (1.9), again, show a transition behavior that is, however, less steep due to the higher critical angle employed for the slip directions (b). Contrary to the case of rotation (a), (6.3) is not affected by the rotation and is, therefore, constant. This appears to be unphysical and is caused by the neglect of the slip directions in this factor. For the employed rotation of the slip direction, (6.1) approaches infinite resistance of the GB against transmission of plastic slip at $\varphi = 90°$.

For a rotation of both slip direction and slip plane normal about an axis perpendicular to them (Fig. 6.1c), (1.4) predicts optimal transmission at

$\varphi = 90°$. This appears to be unphysical since both slip direction and slip plane normal in grain B are perpendicular to their counterparts in grain A which would severely restrict transmission. The factors (1.5)–(1.6), however, predict no transmission at this angle. Factor (1.7) is constant for this case since the lines of intersection do not change for the employed rotation. The depicted jump is of numerical nature and does not alter the qualitative behavior. Factor (1.8) considers two critical angles, and thus, in this case, the smaller one of both angles limits the transmission regime. Although (1.9) also takes into account two critical angles, the larger one limits the transmission in the investigated case. This is due to the consideration of the angle between the lines of intersection (which does not change in this case) and the neglect of slip directions. The product (1.11a) shows a slightly larger regime of no transmission, compared to the previous two criteria. For (6.3), the critical angle to be considered is the angle between the slip directions since the lines of intersection are unaltered. This gives a slightly larger regime of possible plastic slip transmission than for rotation (a). In this context, it is also questionable why limit values should be employed for both the lines of intersection and the slip plane normals. The angle between slip plane normals is always larger than the angle between the lines of intersection (see Werner and Prantl, 1990). The criterion (6.1) shows identical behavior as in the previous case. This is also questionable since the alignment of slip planes has been found to be more substantial in predicting transmission of plastic slip and, thus, would be expected to limit transmissibility upon applying an identical rotation to the slip direction and the slip plane normal.

In the last case, the GB is rotated about an axis perpendicular to the slip directions and the slip plane normals of both grains (Fig. 6.1d). The slip systems, however, are not altered, at first. This case, thus, isolates the influence of the GB when comparing the different GTFs. Interestingly, the orientation of the GB does not influence any of the investigated factors for this case (see dashed lines in Fig. 6.2d).

Then, the slip system in grain B is pre-rotated by $15°$ about an axis $\boldsymbol{a}$ (Fig. 6.1d). The GB is rotated again about the depicted axis $\boldsymbol{r}$ (Fig. 6.1d). Both factors (1.4) and (1.5) are not affected by this rotation since they do not account for the GB orientation. Instead, constant results are obtained for this case (and slightly smaller than "one" since the slip directions are changed, initially). The factor (1.8) does also not account for the GB orientation and shows a constant result. Factors (1.6) and (1.7), however, both account for the GB orientation. The product (1.11a) also considers the orientation of the GB. Factor (1.9) is clearly affected by the GB orientation, and factor (6.3) also shows slight changes upon the employed rotation of the GB. An effective cut-off angle of $\varphi \approx 61°$ is obtained for both (1.9) and (6.3) (see Fig. 6.2d) due to the inherent limiting angle for the lines of intersection alignment. Furthermore, the GB orientation is not considered by (6.1) at all. The slightly higher value than in the previous case is a result of the employed initial rotation for the slip direction in grain B.

6.5 Discussion of the transmission factors in terms of physicality

The comparison of the geometric criteria in this work results in various findings. First of all, the geometric criterion (1.4) contradicts the common understanding of slip transmission since it predicts fully possible transmission for two slip systems with slip directions being perpendicular to each other. The factor (1.5), however, predicts no transmission for this case but neglects the orientation of the GB. Factors (1.6) and (1.7) account for the orientation of the GB and have shown better agreement with experiments if combined with additional considerations regarding the RBV and the RSSes (Lee et al., 1989b; 1990; Clark et al., 1992). Whether the inclination of the GB needs to be considered in such criteria is debatable. In Bachurin et al. (2010), for example, no influence of the

GB inclination on the slip transmission process was found. Indications in this direction are also present in Liu et al. (1995), where the influence of the GB orientation with regard to the loading direction was found to be of minor importance for the dislocation nucleation processes investigated on ice crystals.

Another open question is whether critical angles should be used, e.g., employed in (1.8), (1.9), and (6.3). These criteria predict a far less pronounced regime for possible plastic slip transmission than criteria without such angles. In addition, the experimental foundation for the critical angles is not exhaustive. In fact, some of them (e.g., Werner and Prantl, 1990) have been determined based on an observed range of possible transmission of plastic slip (Davis et al., 1966). Criterion (6.3) does not seem to be applicable for general plastic slip transmission processes since the orientation of the slip directions is not considered which contradicts experimental findings. In the same spirit, (6.1) neglects the slip plane orientations and thus is not applicable when the slip planes for a transmission process are distinct. The use of (weighted) sums of geometric slip transmission factors necessitates further investigations. In Bieler et al. (2014), similar distributions of such weighted approaches were found, although different geometric factors and weights were employed. Furthermore, there are apparent contradictions between the use of (1.8) and experimental findings (Kumar, 2010), also questioning the use of this factor. It would be desirable to compare the geometric transmission criteria including additional criteria (RSS, RBV) for multiple slip systems, preferably also for different crystal structures to gain a deeper understanding of their limitations.

In a recent contribution (Mercier et al., 2015), a Matlab toolbox has been presented that allows to determine the activation of plastic slip based on several of the presented criteria for a bicrystal. This could be used in comparison to experimental results and for the development of physically based continuum approaches for the transfer of plastic slip at GBs. Similarly, the "slip transfer resistance of neighbouring grains"

(STRONG) method by Knorr (2014) can be used for the determination of the possible activity of plastic slip systems near GBs (see, for example, the investigation by Kheradmand et al., 2016).

6.6 Limitations of the gradient crystal plasticity framework with an accumulated plastic slip to consider orientation-dependent grain boundary mechanisms

The presented theory in this thesis is based on the central simplification that an accumulated plastic slip, Eq. (2.6), and its gradient can be used instead of accounting for all plastic slips of the slip systems as DOFs and for their gradients, respectively. Thus, the resulting consideration of only one higher-order stress $\boldsymbol{\xi}$ limits the model ability to account for effects by slip system interactions. While this choice is of substantial benefit for the numerical implementation due to the significant reduction of DOFs in the model, it also leads to some limitations with regard to the modeling of GB plasticity. Furthermore, it has been assumed that the micromorphic counterpart ζ of γ_{ac} does not feature discontinuities across GBs (see Section 3.2). Therefore, effects stemming from discontinuities of plastic slip across GBs are not expected to be reproducible by the theory at hand.
One possibility to include a crystal-orientation dependent GB yield mechanism would be to postulate the energetic contribution to the GB yield strength as a function of the GB / slip system orientations

$$\Xi_\Gamma^e = \Xi_0^C \left(\varphi_1, \varphi_2, \varphi_3, \varphi_4, \varphi_5\right), \tag{6.12}$$

where the five angles correspond to the macroscopic DOFs (rotations) of a general GB (Wolf, 1990), and the three microscopic DOFs (translations) have been neglected due to the use of a continuum theory. However, the specific form of the right-hand side of Eq. (6.12) is unknown and the parameter-space of this function is comparably vast. This dependence on the GB (mis)orientation and the slip system orientations of both grains would also have to be identified for the GB hardening-relation proposed in Eq. (4.3), i.e., additionally,

$$K_{\mathrm{H}} = K_{\mathrm{H}}\left(\varphi_1, \varphi_2, \varphi_3, \varphi_4, \varphi_5\right). \tag{6.13}$$

A reduction of the complexity of the corresponding identification could be achieved by, for example, limiting the model to consider only the misorientation of the crystal lattices and neglect the orientation of the GB. In this case the GB resistance could be formulated, for instance, as

$$\Xi_\Gamma^{\mathrm{e}} = \Xi_0^{\mathrm{C}}\left(\varphi_{\mathrm{mis}}\right), \tag{6.14}$$

and the GB hardening parameter as

$$K_{\mathrm{H}} = K_{\mathrm{H}}\left(\varphi_{\mathrm{mis}}\right), \tag{6.15}$$

with the minimum misorientation angle φ_{mis}, necessary to rotate one crystal-lattice orientation into the other crystal-lattice orientation around a corresponding axis. From experiments, however, it is well known that the local orientation of the GBs can strongly influence the slip transfer and activation processes (see Section 1.3). Furthermore, at the GBs, effects resulting from residual defects should be accounted for. These are caused, for example, by dislocations leaving residual dislocations within the GBs upon a transmission. Such effects can only be considered if the jumps of plastic slips are evaluated at the GBs.

In the theory of Gurtin (2008), the plastic slips γ_α and the corresponding rates $\dot{\gamma}_\alpha$ are considered. The use of the GB Burgers tensor in combination with a quadratic energy formulation leads to the consideration of all plastic slips and of interactions of all slip systems, as it has been outlined in Section 6.3. For the net-power expended on, for example, a GB by the microtractions, the theory of Gurtin gives

$$\mathcal{P}_\Gamma^\alpha = \int_\Gamma [\![(\boldsymbol{\xi}_\alpha \cdot \boldsymbol{n})\,\dot{\gamma}_\alpha]\!]\,\mathrm{d}a, \tag{6.16}$$

for a slip system α. In comparison, within the accumulated plastic slip framework of the present thesis, the following net-power expended on a GB is obtained, assuming $\zeta = \gamma_{\mathrm{ac}}$,

$$\mathcal{P}_\Gamma = \int_\Gamma [\![\boldsymbol{\xi} \cdot \boldsymbol{n}]\!]\,\dot{\gamma}_{\mathrm{ac}}\,\mathrm{d}a, \qquad \gamma_{\mathrm{ac}} = \sum_\alpha \gamma_\alpha \tag{6.17}$$

(see also the second term of the right-hand side in Eq. (3.2) and the GB microtraction in Box 3.1). Although the net-power in Eq. (6.17) is formally similar to Eq. (6.16), a direct connection between both relations does not appear to be drawable. Therefore, it does also not appear to be possible to connect the GB microtraction of the employed model, $\Xi_\Gamma = [\![\boldsymbol{\xi} \cdot \boldsymbol{n}]\!]$, to the microtractions used in the model by Gurtin. In fact, the introduction of the accumulated plastic slip at the onset of the theory development (Section 3.2), and the consideration of only the gradient of the micromorphic variable, limit the modeling to the consideration of one higher-order stress $\boldsymbol{\xi}$. This approach, unfortunately, rules out a physically sound refinement of the theory to incorporate effects from the interaction of slip systems and their gradient-induced backstresses $\boldsymbol{\xi}_\alpha$.

Concluding, a refinement of the present theory would require to consider both the individual plastic slips as degrees of freedom and the jumps of these quantities at the GBs in conjunction with the jumps of the stresses $\boldsymbol{\xi}_\alpha$ at the GBs. Within such a framework, the orientations of slip systems can naturally be incorporated in the plastic slip mechanisms, as it has been discussed, for example, in Section 6.3.

Chapter 7

Summary and outlook

In this thesis, several questions of continuum modeling of dislocation-based plasticity are addressed with the focus on gradient plasticity modeling. Therefore, the gradient plasticity framework by Wulfinghoff et al. (2013) is taken as the starting point for comparison of finite element simulation results to experimental data of oligocrystalline specimens for different loading conditions. Further development of the gradient plasticity theory is motivated both by the comparison to experiments and to discrete dislocation dynamics simulations, a modeling approach which resolves individual dislocations and their interactions.

An accumulated plastic slip is used within a micromorphic implementation of the theory to keep the arising computational costs (due to the consideration of gradients) comparably low. Using the principle of virtual power, field equations are derived and the boundary conditions are formulated including higher-order boundary conditions which are necessary since the gradient of the micromorphic variable is considered in the theory.

Using the dissipation inequality, the stresses on the slip systems are derived with different contributions from the postulated free energy terms. Work-hardening is considered using a phenomenological Voce-hardening formulation, at first, and a power-law flow rule is used for the bulk material. It is discussed, in detail, that the Voce-hardening formulation in terms of the accumulated plastic slip can be motivated from the evolution law of the dislocation-density in conjunction with

the Taylor-relation assuming constant shear rates and considering only monotonic loading processes.
Using the gradient of the micromorphic variable, instead of all plastic slip gradients, allows to perform fully three-dimensional finite element simulations of different micron-sized grain aggregates. Numerical efficiency of the micromorphic implementation is shown by finite element simulations for several hundreds of thousands degrees of freedom. The following comparisons of simulation results to experiments and discrete simulation results are carried out:

- comparison of simulation results from the basic crystal plasticity framework using an accumulated plastic slip to torsion experiments on bamboo-structured gold microwires
- investigation of the influence of misorientation on the resulting accumulated plastic slip field distributions in comparison to the experimental results
- simulation of size effects which were observed on oligocrystalline copper microwires in tensile test experiments
- simulation of contrary size effects, which were experimentally observed on oligocrystalline gold microwires under both tensile and torsion loading, with consideration of the influence of two distinct crystal-orientations
- comparison of gradient plasticity simulation results to data from discrete dislocation dynamics simulations for an aggregate composed of three grains

In the comparison to experimental results on bamboo-structured gold microwires, it is shown that the overall strain gradient imposed by the torsion loading can be qualitatively reproduced. The experimentally observed variations of the kernel average misorientation are explained by the investigation of small misorientations from the ideal crystal-orientation using the crystal plasticity simulations. Thereby, the feasibility of the used framework to effectively carry out three-dimensional simulations, which aid in the understanding of experimental observations, is demonstrated. However, the experimentally observed annealing of dislocations close to the surface of the bamboo-structured microwires cannot be accounted for within the present theory. A physically refined theory would be required to relax the restriction of using a (non-decreasing) accumulated plastic slip to model the plastic response.

The overall size effect on copper microwires under tensile loading with different grain sizes is reproduced using the gradient plasticity framework with a constant value of the grain boundary yield strength. In a subsequent comparison to experiments on gold microwires of different diameters and grain sizes, the observed inverse size effect under tensile loading and the "classic" size effect under torsion loading is explained, considering the influence of the texture of the specimens in a simplified fashion. The grain boundary yield strength, however, is adjusted to the individual microwires due to discrepancies in the average grain sizes limited by the computational modeling. In addition, the magnitude of the overall size effect under torsion loading is substantially underestimated when using the model parameters identified from the tensile loading case. This might be indicating that the used framework could be limited when comparably inhomogeneous deformations (like torsion loading) are applied. In the comparison to discrete dislocation dynamics simulations, it is shown that the sole use of the Voce-hardening relation in the gradient plasticity model leads to an overestimation of the plastic strain near grain boundaries, as the evolution of plastic strain, there, cannot be modeled appropriately.

Motivated by this comparison, enhancements of the theoretical framework and the numerical implementation as well as the following comparisons are performed:

- extension of the grain boundary yield condition by an additional grain boundary hardening relation

- comparison of the gradient plasticity simulation results with grain boundary hardening to results from discrete dislocation dynamics simulations
- replacement of the quadratic defect energy formulation by a generalized form of power-law type
- numerical regularization of the higher-order stress for vanishing gradients of the micromorphic variable
- exact solution for single slip of a laminate using the power-law defect energy
- investigation of gradient plasticity simulation results in dependence of the power-law defect energy exponent and the associated internal length scale, accompanied by results of an exact solution
- comparison of gradient plasticity simulation results using different defect energy power-law exponents to discrete dislocation dynamics results

Using grain boundary hardening, the evolution of plastic strain near the grain boundaries is modeled substantially better, compared to using a constant grain boundary yield strength and Voce-hardening for the bulk material. The differences in the plastic strain distributions of the grains due to different grain misorientations, however, could not be modeled. Gradient plasticity simulations using either Voce-hardening or grain boundary hardening are considered but a combination of both is not investigated due to the lack of a suitable benchmark problem for calibration. With the, additionally, considered defect energy of power-law type, different distributions of plastic strain can be modeled since the choice of the exponent alters the shape of the distributions. Therefore, the observable differences in the gradients of plastic strain near the grain boundaries are reduced by the employment of the power-law defect energy. It is discussed, however, that the current model restricts the development of the gradients near the grain boundaries due to the limitation to one higher-order vectorial gradient-stress. This prevents

an even better agreement of gradient plasticity simulation results with the discrete dislocation dynamics simulations results.
The influence of the misorientation of the crystal-orientations from adjacent grains as well as of the grain boundary orientation is not considered in the present grain boundary yield condition. Therefore, the following points are addressed in the present work as a contribution towards further refinement of the model:

- an overview is given on the dislocation mechanisms and interactions at grain boundaries
- an overview is given on the literature on slip transmission criteria that are used both in experiments and in continuum simulations
- a connection is drawn between these criteria and a model that naturally incorporates effects from the orientation-dependent interaction of slip systems (Gurtin, 2008)
- the geometrical criteria for slip transmission are compared for the single slip case in order to gain a basic understanding of the physical limitations of the geometric criteria from the literature
- the limitations of the employed grain boundary model are discussed with regard to a possible consideration of the orientation-dependency in the GB yield condition

Additionally, in the comparison to discrete dislocation dynamics simulations, it is shown that the orientation-dependent interactions of dislocations across grain boundaries are not resolved in the current continuum model. At the same time, the overall size effect of the considered gold microwire torsion experiments could not be modeled quantitatively. Therefore, several questions remain open that should be addressed in future works. In detail, these are:

- consideration of a suitable benchmark problem for an investigation of the model using both Voce-hardening of the bulk and additional grain boundary hardening

- resolving the limitation that the development of gradients of plastic strain near the grain boundaries is restricted due to the consideration of only one higher-order vectorial gradient-stress
- the consideration of orientation-dependency in the grain boundary yield condition or using a model considering all plastic slips and their gradients
- repetition of the modeling-attempt to reproduce the overall size effect of gold microwires under torsion loading with a physically more advanced model
- modeling of the elastic interactions of dislocations across grain boundaries that result in crystal-orientation dependent field distributions of plastic strain
- comparison of simulation results from the employed gradient plasticity model to results from physically more sophisticated theories (e.g., Gurtin, 2008)
- extended comparison of gradient plasticity simulation results to experimental results featuring inhomogeneous deformations
- a generalization of the theory to finite deformations

Appendix A

Overview of slip transmission criteria from the literature

Table A.1: Slip transmission criteria in the experimental literature. Abbreviations used are AC: Additional criteria, REF: Reference, RSS: Resolved shear stress / Schmid factors, RBV: Residual Burgers vector, PKF: Peach-Koehler force, IBS: Interface barrier strength, SIF: Stress intensity factor, SW: Slip weights, SFW: Schmid factor weights, GTF: Geometrical transmission factor. Table reprinted from Bayerschen et al. (2016a) with permission from Springer.

GTF	AC REF	GTF	AC REF		
$\hat{N}_{\alpha\beta}$	- [1], [2]	$\hat{N}^{\mathrm{mod}}_{\alpha\beta}$	- [3], [4]	RSS / RBV / SIF [5]	RSS [6]
$\hat{M}^{\mathrm{mod}}_{\alpha\beta}$	RSS / RBV [7]-[11]	$\hat{M}_{\alpha\beta}$	- [12]	PKF [13], [14]	RBV [15]
$\hat{\chi}_{\alpha\beta}$	RSS / IBS [16]	$\bar{\lambda}$	- [17]	RSS [18]	
m'_{γ}, m'_{m}	SW, SFW [6]	LRB_{γ}, s_{γ}	SW, SW [6]		

[1] Livingston and Chalmers (1957)
[2] Davis et al. (1966)
[3] Luster and Morris (1995)
[4] Wo and Ngan (2004)
[5] Guo et al. (2014)
[6] Bieler et al. (2014)
[7] Clark et al. (1992)
[8] Lee et al. (1989b)
[9] Lee et al. (1990)
[10] Misra and Gibala (1999)
[11] Abuzaid et al. (2012)
[12] Soer and De Hosson (2005)
[13] Shen et al. (1986)
[14] Shen et al. (1988)
[15] Tiba et al. (2015)
[16] Beyerlein et al. (2012)
[17] Werner and Prantl (1990)
[18] Kumar (2010)

Table A.2: Slip transmission / interaction criteria in crystal plasticity models. Abbreviations are AC: Additional criteria, REF: Reference, RSS: Resolved shear stress / Schmid factors, RBV: Residual Burgers vector, GTF: Geometrical transmission factor. The approaches by Gurtin (2008); Özdemir and Yalçinkaya (2014) utilize $\hat{M}_{\alpha\beta}$ rather as an *inter*-action coefficient than as a classic transmission factor. Table reprinted from Bayerschen et al. (2016a) with permission from Springer.

GTF	$\tan(\varphi_\alpha^{\mathrm{AB}})$	$\hat{M}_{\alpha\beta}$	$\hat{M}_{\alpha\beta}$	-	$\hat{\zeta}_{\alpha\beta}$	$\hat{\chi}_{\alpha\beta}$
AC	RSS	RSS	RBV / RSS	RBV / RSS	RSS	RSS
REF	[19]	[20], [21]	[22], [23]	[24]	[25]	[26]

[19] Ekh et al. (2011)
[20] Shi and Zikry (2009)
[21] Shi and Zikry (2011)
[22] Gurtin (2008)
[23] Özdemir and Yalçinkaya (2014)
[24] Ma et al. (2006)
[25] Ashmawi and Zikry (2002)
[26] Mayeur et al. (2015)

Appendix B

Theorems

In the three-dimensional space, with position-dependent quantities $\boldsymbol{A}, \boldsymbol{b}, c$, the divergence theorem can be expressed by the following forms

$$\int_V \boldsymbol{A} \cdot \operatorname{grad}(\boldsymbol{b}) \, \mathrm{d}v = -\int_V \operatorname{div}(\boldsymbol{A}) \cdot \boldsymbol{b} \, \mathrm{d}v + \int_{\partial V} \boldsymbol{A}\boldsymbol{n} \cdot \boldsymbol{b} \, \mathrm{d}a, \tag{B.1}$$

$$\int_V \boldsymbol{a} \cdot \operatorname{grad}(c) \, \mathrm{d}v = -\int_V \operatorname{div}(\boldsymbol{a}) \, c \, \mathrm{d}v + \int_{\partial V} (\boldsymbol{a} \cdot \boldsymbol{n}) \, c \, \mathrm{d}a, \tag{B.2}$$

$$\int_V \boldsymbol{a} \cdot \operatorname{grad}(c) \, \mathrm{d}v = -\int_V \operatorname{div}(\boldsymbol{a}) \, c \, \mathrm{d}v + \int_{\partial V} (\boldsymbol{a} \cdot \boldsymbol{n}) \, c \, \mathrm{d}a - \int_{\Gamma} [\![\boldsymbol{a}]\!] \cdot \boldsymbol{n} c \, \mathrm{d}a, \tag{B.3}$$

where a material volume V with outer surface ∂V and interface Γ has been considered. Jump contributions are neglected in (B.1) and (B.2).

Appendix C

Micromorphic approximation of the accumulated plastic slip

C.1 Single-crystalline case

Exemplary, the two field distributions γ_{ac} (Fig. C.1.1a) and ζ (Fig. C.1.1b) are compared for the top cross-section of the single-crystalline simulations from Fig. 2.7 at the final time step. No pronounced differences between both distributions are observable. If a sufficiently fine mesh and a large enough penalty-parameter H_χ are used both distributions can be regarded as approximately the same. Therefore, the micromorphic concept can be used and the approximation of the accumulated plastic slip γ_{ac} is obtained by ζ.

C.2 Oligocrystalline case

For the case of considering grain boundaries in the model, a comparison of both quantities (ζ, γ_{ac}) has been performed in Wulfinghoff et al. (2013). The distributions of ζ and γ_{ac} are depicted in Figure C.2.1 for a 64-grain-aggregate. Small local deviations occur for the selected value of the penalty parameter. However, the overall field solutions of both quantities are still in fair agreement and the approximation of γ_{ac} by ζ can be used.

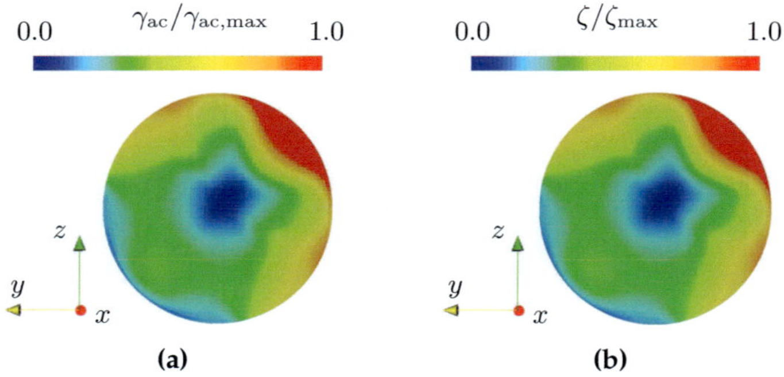

Figure C.1.1: (a) Normalized field distribution of γ_{ac} in top cross-section at final time step of simulation. (b) Normalized field distribution of ζ in top cross-section at final time step of simulation for $\zeta_{max} = \gamma_{ac,max}$.

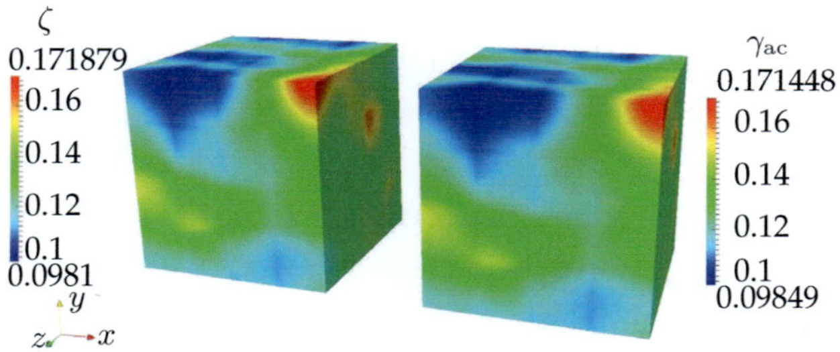

Figure C.2.1: Resulting distributions of the accumulated plastic slip γ_{ac} and its micromorphic counterpart ζ for a tensile test simulation with 64 grains. Figure reprinted from Wulfinghoff et al. (2013) with permission from Elsevier.

Appendix D

Computational times of the finite element implementation

Table D.1: Exemplary computational times of finite element simulations for different values of the parameter Ξ_0^C, where "0" corresponds to a simulation without grain boundary resistance. Calculations are performed on an Intel Xeon CPU E5540 2.53GHz. The references to the corresponding overall mechanical responses are given in the last column of the table.

Loading case	Grains	Ξ_0^C in N/m	Number time steps	Comp. time in s	DOFs	Fig.
Tensile	1	0	11	3239	160992	2.9a
Torsion	1	0	11	8343	160992	2.9b
Tensile	3	∞	15	19541	349804	5.3a
Tensile	64	0	12	18042	275684	3.3
Tensile	64	50	10	17112	275684	3.3
Tensile	64	100	15	28786	275684	3.3
Tensile	64	11	21	74118	235396	3.10a
Torsion	64	11	13	37780	235396	3.10b

Appendix E

Convention for slip systems of face-centered cubic unit cell

E.1 Slip systems in the $\langle 100 \rangle$-orientation

The slip system convention used for the implementation of the FCC materials is listed below. Slip directions are clustered by the corresponding slip plane normals and are listed in the unrotated initial orientation. In this orientation, the edges of the unit cell are aligned with the axes of the Cartesian coordinate system such that the $\langle 100 \rangle$-crystal-direction is parallel to the tensile axis or the axis of rotation in the case of torsion loading, respectively. The net-plastic-slip of a slip system α is obtained by $\gamma_\alpha = \lambda_\alpha - \lambda_{\alpha+12}$ and the accumulated plastic slip by $\tilde{\gamma}_\alpha = \lambda_\alpha + \lambda_{\alpha+12}$.

$$\alpha = 1: \quad \boldsymbol{d}_1 = \frac{1}{\sqrt{2}} \begin{pmatrix} 0 \\ 1 \\ -1 \end{pmatrix}, \quad \bar{\boldsymbol{d}}_1 = \frac{1}{\sqrt{2}} \begin{pmatrix} 0 \\ -1 \\ 1 \end{pmatrix}, \quad \boldsymbol{n}_{1,2,3} = \frac{1}{\sqrt{3}} \begin{pmatrix} 1 \\ 1 \\ 1 \end{pmatrix}$$

$$\alpha = 2: \quad \boldsymbol{d}_2 = \frac{1}{\sqrt{2}} \begin{pmatrix} -1 \\ 0 \\ 1 \end{pmatrix}, \quad \bar{\boldsymbol{d}}_2 = \frac{1}{\sqrt{2}} \begin{pmatrix} 1 \\ 0 \\ -1 \end{pmatrix}, \quad \boldsymbol{n}_{1,2,3} = \frac{1}{\sqrt{3}} \begin{pmatrix} 1 \\ 1 \\ 1 \end{pmatrix}$$

$$\alpha = 3: \quad \boldsymbol{d}_2 = \frac{1}{\sqrt{2}} \begin{pmatrix} 1 \\ -1 \\ 0 \end{pmatrix}, \quad \bar{\boldsymbol{d}}_3 = \frac{1}{\sqrt{2}} \begin{pmatrix} -1 \\ 1 \\ 0 \end{pmatrix}, \quad \boldsymbol{n}_{1,2,3} = \frac{1}{\sqrt{3}} \begin{pmatrix} 1 \\ 1 \\ 1 \end{pmatrix}$$

$$\alpha = 4: \quad \boldsymbol{d}_4 = \frac{1}{\sqrt{2}}\begin{pmatrix} 0 \\ 1 \\ -1 \end{pmatrix}, \quad \bar{\boldsymbol{d}}_4 = \frac{1}{\sqrt{2}}\begin{pmatrix} 0 \\ -1 \\ 1 \end{pmatrix}, \quad \boldsymbol{n}_{4,5,6} = \frac{1}{\sqrt{3}}\begin{pmatrix} -1 \\ 1 \\ 1 \end{pmatrix}$$

$$\alpha = 5: \quad \boldsymbol{d}_5 = \frac{1}{\sqrt{2}}\begin{pmatrix} 1 \\ 0 \\ 1 \end{pmatrix}, \quad \bar{\boldsymbol{d}}_5 = \frac{1}{\sqrt{2}}\begin{pmatrix} -1 \\ 0 \\ -1 \end{pmatrix}, \quad \boldsymbol{n}_{4,5,6} = \frac{1}{\sqrt{3}}\begin{pmatrix} -1 \\ 1 \\ 1 \end{pmatrix}$$

$$\alpha = 6: \quad \boldsymbol{d}_6 = \frac{1}{\sqrt{2}}\begin{pmatrix} -1 \\ -1 \\ 0 \end{pmatrix}, \quad \bar{\boldsymbol{d}}_6 = \frac{1}{\sqrt{2}}\begin{pmatrix} 1 \\ 1 \\ 0 \end{pmatrix}, \quad \boldsymbol{n}_{4,5,6} = \frac{1}{\sqrt{3}}\begin{pmatrix} -1 \\ 1 \\ 1 \end{pmatrix}$$

$$\alpha = 7: \quad \boldsymbol{d}_7 = \frac{1}{\sqrt{2}}\begin{pmatrix} 0 \\ 1 \\ 1 \end{pmatrix}, \quad \bar{\boldsymbol{d}}_7 = \frac{1}{\sqrt{2}}\begin{pmatrix} 0 \\ -1 \\ -1 \end{pmatrix}, \quad \boldsymbol{n}_{7,8,9} = \frac{1}{\sqrt{3}}\begin{pmatrix} -1 \\ 1 \\ -1 \end{pmatrix}$$

$$\alpha = 8: \quad \boldsymbol{d}_8 = \frac{1}{\sqrt{2}}\begin{pmatrix} 1 \\ 0 \\ -1 \end{pmatrix}, \quad \bar{\boldsymbol{d}}_8 = \frac{1}{\sqrt{2}}\begin{pmatrix} -1 \\ 0 \\ 1 \end{pmatrix}, \quad \boldsymbol{n}_{7,8,9} = \frac{1}{\sqrt{3}}\begin{pmatrix} -1 \\ 1 \\ -1 \end{pmatrix}$$

$$\alpha = 9: \quad \boldsymbol{d}_9 = \frac{1}{\sqrt{2}}\begin{pmatrix} -1 \\ -1 \\ 0 \end{pmatrix}, \quad \bar{\boldsymbol{d}}_9 = \frac{1}{\sqrt{2}}\begin{pmatrix} 1 \\ 1 \\ 0 \end{pmatrix}, \quad \boldsymbol{n}_{7,8,9} = \frac{1}{\sqrt{3}}\begin{pmatrix} -1 \\ 1 \\ -1 \end{pmatrix}$$

$$\alpha = 10: \quad \boldsymbol{d}_{10} = \frac{1}{\sqrt{2}}\begin{pmatrix} 0 \\ 1 \\ 1 \end{pmatrix}, \quad \bar{\boldsymbol{d}}_{10} = \frac{1}{\sqrt{2}}\begin{pmatrix} 0 \\ -1 \\ -1 \end{pmatrix}, \quad \boldsymbol{n}_{10-12} = \frac{1}{\sqrt{3}}\begin{pmatrix} 1 \\ 1 \\ -1 \end{pmatrix}$$

$$\alpha = 11: \quad \boldsymbol{d}_{11} = \frac{1}{\sqrt{2}}\begin{pmatrix} 1 \\ -1 \\ 0 \end{pmatrix}, \bar{\boldsymbol{d}}_{11} = \frac{1}{\sqrt{2}}\begin{pmatrix} -1 \\ 1 \\ 0 \end{pmatrix}, \boldsymbol{n}_{10-12} = \frac{1}{\sqrt{3}}\begin{pmatrix} 1 \\ 1 \\ -1 \end{pmatrix}$$

$$\alpha = 12: \quad \boldsymbol{d}_{12} = \frac{1}{\sqrt{2}}\begin{pmatrix} 1 \\ 0 \\ 1 \end{pmatrix}, \bar{\boldsymbol{d}}_{12} = \frac{1}{\sqrt{2}}\begin{pmatrix} -1 \\ 0 \\ -1 \end{pmatrix}, \boldsymbol{n}_{10-12} = \frac{1}{\sqrt{3}}\begin{pmatrix} 1 \\ 1 \\ -1 \end{pmatrix}$$

E.2 Slip systems in the $\langle 111 \rangle$-orientation

In order to rotate the crystal-orientations of the grains and, thus, their slip directions and normals into the $\langle 111 \rangle$-orientation, a rotation-matrix in z-x-z-convention (Bunge, 2013) is used. The rotation of an exemplary vector $\boldsymbol{a}$ is given by

$$\bar{\boldsymbol{a}} = \boldsymbol{Q}\boldsymbol{a}, \tag{E.1}$$

with the orthogonal rotation matrix

$$\boldsymbol{Q} = \boldsymbol{Q}_1\boldsymbol{Q}_2\boldsymbol{Q}_3, \tag{E.2}$$

where the three rotation matrices are

$$\boldsymbol{Q}_1 \mathrel{\widehat{=}} \begin{pmatrix} \cos\vartheta_1 & -\sin\vartheta_1 & 0 \\ \sin\vartheta_1 & \cos\vartheta_1 & 0 \\ 0 & 0 & 1 \end{pmatrix}, \tag{E.3}$$

$$\boldsymbol{Q}_2 \mathrel{\widehat{=}} \begin{pmatrix} 1 & 0 & 0 \\ 0 & \cos\vartheta_2 & -\sin\vartheta_2 \\ 0 & \sin\vartheta_2 & \cos\vartheta_2 \end{pmatrix}, \tag{E.4}$$

$$\boldsymbol{Q}_3 \mathrel{\widehat{=}} \begin{pmatrix} \cos\vartheta_3 & -\sin\vartheta_3 & 0 \\ \sin\vartheta_3 & \cos\vartheta_3 & 0 \\ 0 & 0 & 1 \end{pmatrix}. \tag{E.5}$$

The necessary Euler angles to transform the $\langle 100 \rangle$-crystal-orientation into the $\langle 111 \rangle$-crystal-orientation are listed in Table E.2.1 (Jöchen, 2011). To obtain the $\langle 111 \rangle$-orientation, for example, to be aligned parallel with the loading direction of a tensile test simulation, the transposed rotation matrix is used to transform the slip directions and normals via

$$\boldsymbol{Q}^{\mathsf{T}}\bar{\boldsymbol{a}} = \boldsymbol{a}, \tag{E.6}$$

where the identity $\boldsymbol{Q}^{\mathsf{T}} = \boldsymbol{Q}^{-1}$ holds since $\boldsymbol{Q} \in Orth$.

Table E.2.1: Euler angles for rotation of two crystal-orientations of FCC-crystals in used convention to be aligned with the loading axis.

Crystal orientation	Euler angles		
	ϑ_1	ϑ_2	ϑ_3
$\langle 100 \rangle$	0	0	0
$\langle 111 \rangle$	0.6317823	1.357809	0.6322054

Using Eq. (E.6) with the Euler angles from Table E.2.1, for the slip systems $\alpha = 7 - 9$ from Appendix E.1 yields

$$\alpha = 7: \quad \boldsymbol{d}_7 = \sqrt{2}\begin{pmatrix} 0 \\ 0.5 \\ -0.5 \end{pmatrix}, \quad \bar{\boldsymbol{d}}_7 = \sqrt{2}\begin{pmatrix} 0 \\ -0.5 \\ 0.5 \end{pmatrix}, \quad \boldsymbol{n}_{7,8,9} = \begin{pmatrix} 1 \\ 0 \\ 0 \end{pmatrix}$$

$$\alpha = 8: \quad \boldsymbol{d}_8 = \sqrt{2}\begin{pmatrix} 0 \\ 0.183 \\ 0.683 \end{pmatrix}, \quad \bar{\boldsymbol{d}}_8 = \sqrt{2}\begin{pmatrix} 0 \\ -0.183 \\ -0.683 \end{pmatrix}, \quad \boldsymbol{n}_{7,8,9} = \begin{pmatrix} 1 \\ 0 \\ 0 \end{pmatrix}$$

$$\alpha = 9: \quad \boldsymbol{d}_9 = \sqrt{2}\begin{pmatrix} 0 \\ -0.683 \\ -0.183 \end{pmatrix}, \quad \bar{\boldsymbol{d}}_9 = \sqrt{2}\begin{pmatrix} 0 \\ 0.683 \\ 0.183 \end{pmatrix}, \quad \boldsymbol{n}_{7,8,9} = \begin{pmatrix} 1 \\ 0 \\ 0 \end{pmatrix}$$

Appendix F

Convergence studies

F.1 Oligocrystal tensile test simulation using quadratic defect energy

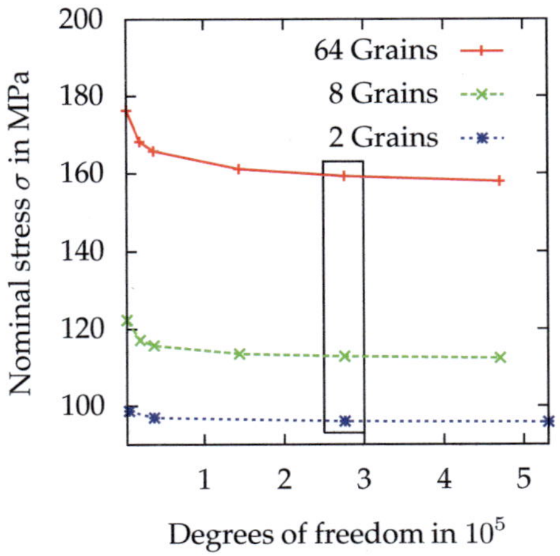

Figure F.1.1: Final nominal stress at the end of the last time step versus degrees of freedom for oligocrystals under tensile loading with 64, 8 and 2 grains, respectively, using different discretizations. The diminution of DOFs due to the Dirichlet boundary conditions is neglected. The chosen discretizations for the simulations are highlighted by a rectangle. Basic figure reprinted from Wulfinghoff and Böhlke (2013) with permission from Elsevier.

F.2 Tricrystal tensile test simulation using power-law defect energy

The convergence of the numerical results for a tricrystal is discussed twofold, here. Besides convergence with regard to the spatial discretization of the finite element mesh, the convergence with respect to the regularization parameter ϵ is investigated. The defect energy exponent used in this section is $m = 1.1$, and the normalization constant is $g_0 = 450.21/\mu\text{m}$. As it can be seen in Fig. F.2.1a, the regularization parameter ϵ is of negligible influence on the final nominal stress at the end of the loading. It also does not influence the convergence behavior regarding the spatial discretization. In Fig. F.2.1b, it is shown that a sufficiently small choice of the regularization parameter leads to coinciding distributions of ζ along the x-axis of the central grain.

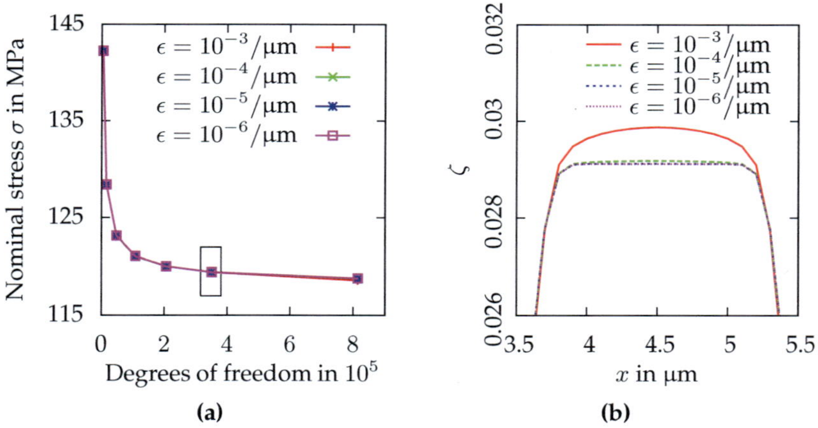

Figure F.2.1: (a) Convergence study: power-law defect energy with $m = 1.1$, $g_0 = 450.21/\mu\text{m}$. Final nominal stress at $t = 1.0$ s vs. DOFs for tricrystal with elastic boundary grains, different spatial discretizations, and different numerical regularization parameter values ϵ. The chosen discretization (349804 DOF) for the simulations is indicated by a rectangle. (b) Distribution of micromorphic variable ζ in central grain for this discretization and different numerical regularization parameter values ϵ. Figures reprinted from Bayerschen and Böhlke (2016) with permission from Springer.

Appendix G

Parameter study of grain boundary hardening

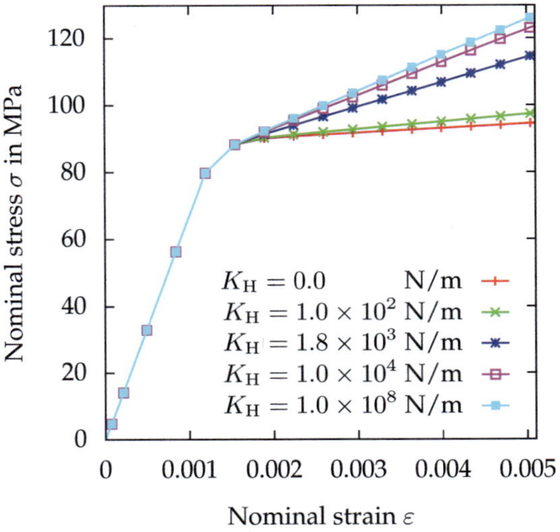

Figure G.1: Parameter study of grain boundary hardening parameter K_H. Case NLC35G, only K_H is varied as shown above. All other parameters can be found in Table 4.5 and Section 4.3.3, respectively. Figure reprinted from Bayerschen et al. (2015).

Appendix H

Derivation of the gradient plasticity field equations without grain boundary contributions

Application of the principle of virtual power $\delta\mathcal{P}_{\rm int} = \delta\mathcal{P}_{\rm ext}$, i.e., letting the right-hand side of Eq. (5.1) being equal to the right-hand side of Eq. (5.2), yields

$$\int\limits_{\mathcal{B}} \left(\boldsymbol{\sigma} \cdot \delta\dot{\boldsymbol{\varepsilon}} + \pi\delta\dot{\zeta} + \boldsymbol{\xi} \cdot \nabla\delta\dot{\zeta}\right) \,\mathrm{d}v = \int\limits_{\partial\mathcal{B}_{\rm t}} \bar{\boldsymbol{t}} \cdot \delta\dot{\boldsymbol{u}} \,\mathrm{d}a + \int\limits_{\partial\mathcal{B}_{\Xi}} \bar{\Xi}\,\delta\dot{\zeta}\,\mathrm{d}a. \tag{H.1}$$

Then, by substituting $\boldsymbol{\varepsilon} = \mathrm{sym}(\mathrm{grad}\,(\boldsymbol{u}))$ in Eq. (H.1) and applying two forms of the divergence theorem, Eq. (B.1) and Eq. (B.3), neglecting any jump contributions, the following is obtained, after regrouping of the terms,

$$-\int\limits_{\mathcal{B}} \mathrm{div}\,(\boldsymbol{\sigma}) \cdot \delta\dot{\boldsymbol{u}}\,\mathrm{d}v + \int\limits_{\partial\mathcal{B}_{\rm t}} (\boldsymbol{\sigma}\boldsymbol{n} - \bar{\boldsymbol{t}}) \cdot \delta\dot{\boldsymbol{u}}\,\mathrm{d}a + \int\limits_{\mathcal{B}} (\pi - \mathrm{div}\,(\boldsymbol{\xi}))\,\delta\dot{\zeta}\,\mathrm{d}v$$

$$+ \int\limits_{\partial\mathcal{B}_{\Xi}} (\boldsymbol{\xi} \cdot \boldsymbol{n} - \bar{\Xi})\delta\dot{\zeta}\,\mathrm{d}a = 0. \tag{H.2}$$

Requiring the left-hand side of Eq. (H.2) to vanish for arbitrary virtual rates $\delta\dot{\boldsymbol{u}}, \delta\dot{\zeta}$ yields the field equations in Table 5.1.

Frequently used acronyms, symbols, and operators

Acronyms

AC	Additional criteria
AT	Annealing temperature
BC	Boundary condition
CP	Crystal plasticity
DDD	Discrete dislocation dynamics
DOF	Degree of freedom
EBSD	Electron backscatter diffraction
ECD	Equivalent circle diameter
FCC	Face-centered cubic
FE	Finite element
FEM	Finite element method
GB	Grain boundary
GND	Geometrically necessary dislocation
GP	Gradient plasticity
GS	Grain size
GTF	Geometric transmission factor
KAM	Kernel average misorientation
KIT	Karlsruhe Institute of Technology
LC	Lateral contraction
LD	Loading direction
NLC	No lateral contraction

OL	Original legend
PKF	Peach-Koehler force
POVP	Principle of virtual power
RBV	Residual burgers vector
REF	Reference
RSS	Resolved shear stress
SFW	Schmid factor weights
SIF	Stress intensity factor
SL	Simplified legend
SSD	Statistically stored dislocation
SW	Slip weights

Latin letters

$a, b, A, B, \mathcal{D}, \ldots$	Scalar quantities
$\boldsymbol{a}, \boldsymbol{b}, \boldsymbol{c}, \ldots$	First-order tensors
$\boldsymbol{A}, \boldsymbol{B}, \boldsymbol{C}, \ldots$	Second-order tensors
$\mathbb{A}, \mathbb{B}, \mathbb{C}, \ldots$	Fourth-order tensors
$\hat{A}, \hat{B}$	Matrices
$\mathcal{A}$	Set of plastically active grain boundary nodes
A_α	Swept area of dislocations of slip system α
b	Burgers vector magnitude
$\boldsymbol{b}$	Burgers vector
$\boldsymbol{b}_{\mathrm{r}}$	Residual Burgers vector
$\mathcal{B}$	Body / material region $\mathcal{B}$
$C_{\alpha\beta}^{\mathrm{AA}}$	Intra-action factors of slip systems in grain A (Gurtin, 2008)
$C_{\alpha\beta}^{\mathrm{AB}}$	Inter-action factors of slip systems in grain A and grain B (Gurtin, 2008)
$C_{1111}, C_{1122}, C_{1212}$	Elastic constants
$\mathbb{C}$	Fourth-order elastic stiffness tensor

d_{avg}	Average grain size
$\boldsymbol{d}_{\alpha}^{\mathrm{A}}$	Slip direction of system α in grain A
$\partial\mathcal{B}$	Boundary of a region $\mathcal{B}$
$\partial\mathcal{B}_{\mathrm{t}}, \partial\mathcal{B}_{\Xi}$	Neumann boundary
$\partial\mathcal{B}_{\mathrm{u}}$	Dirichlet boundary
D	Nominal diameter of microwire
D_{exp}	Real diameter of microwire
D_{sim}	Diameter of cylindrical simulation volume
D_{tot}	Total dissipation
$\mathcal{D}$	Dissipation density
$\mathcal{D}_{\Gamma}$	Grain boundary dissipation density
$\boldsymbol{e}_{x}$	Unit vector, e.g., in x-direction of a Cartesian coordinate system
E	Young's modulus
f_{Γ}	Grain boundary yield function
g_{0}	Normalization constant of defect energy density
G	Shear modulus
$\boldsymbol{G}$	Grain boundary Burgers tensor (Gurtin, 2008)
$\boldsymbol{H}$	Displacement gradient
$\boldsymbol{H}^{\mathrm{e}}$	Elastic distortion
$\boldsymbol{H}^{\mathrm{p}}$	Plastic distortion
H_{χ}	Numerical penalty parameter
$\mathbb{I}^{\mathrm{s}}$	Fourth-order symmetric identity tensor
$\boldsymbol{I}$	Second-order identity tensor
k_{1}, k_{2}	Parameters of dislocation density evolution law
K_{G}	Defect energy parameter (alternative formulation)
K_{H}	Grain boundary hardening parameter
$\boldsymbol{l}_{\alpha}^{\mathrm{A}}$	Line of intersection vector of system α in grain A with grain boundary
l, L	Edge length of grain discretization
$LRB_{\gamma}, m'_{m}, m'_{\gamma}, s_{\gamma}$	Weighted sum criteria (Bieler et al., 2014)

l_{int}	Internal length scale
$\tilde{\boldsymbol{l}}$	Dislocation line direction
L_0	Length of simulation volume
$\hat{M}_{\alpha\beta}^{\text{mod}}$	Geometric slip transmission matrix (Lee et al., 1989b)
$\hat{M}_{\alpha\beta}$	Geometric slip transmission matrix (Shen et al., 1986)
m	Exponent of power-law defect energy
m_{α}^{A}	Schmid factor of system α in grain A
M_T	Torque
n	Time step number
$\boldsymbol{n}_{\Gamma}$	Grain boundary normal
$\boldsymbol{n}_{\alpha}^{\text{A}}$	Slip plane normal of system α in grain A
$\hat{N}_{\alpha\beta}$	Geometric slip transmission matrix (Livingston and Chalmers, 1957)
$\hat{N}_{\alpha\beta}^{\text{mod}}$	Geometric slip transmission matrix (Luster and Morris, 1995)
$Orth$	Set of orthogonal tensors / matrices
p	Rate sensitivity parameter
p_{vol}	Power density of bulk material
$\check{p}$	Numerical penalty stress
P_{ext}	External power
P_{int}	Internal power
$\mathcal{P}_{\Gamma}$	Net power expended on grain boundary
$\boldsymbol{Q}, \boldsymbol{Q}_1, \boldsymbol{Q}_2, \boldsymbol{Q}_3$	Orthogonal rotation matrices
r	Radius to central axis of a cylindrical volume
$r^{\text{p}}, \boldsymbol{r}^{\sigma}$	Residuals
res	Maximum norm of force residual
res_0	Initial residual
R	Maximum radius of a cylindrical volume
s	Scaling factor
$\tilde{s}$	Parametrization variable

Sym	Set of symmetric tensors / matrices
t	Time
$\mathrm{tol}_{\mathcal{A}}$	Tolerance for active-set search
$\mathrm{tol}_{\mathrm{force}}$	Tolerance for force residual
$\boldsymbol{t}$	Traction / stress vector
$\boldsymbol{u}$	Displacement vector
$\boldsymbol{u}^{\mathrm{e}}$	Elastic displacement vector
V	Volume
w_i	Integration point weights
W	Free energy density
W_{e}	Elastic contribution to the free energy density
W_{Γ}	Free energy density of the grain boundaries
$\tilde{W}_{\mathrm{g}}$	Numerically regularized gradient contribution to the free energy density
W_{g}	Gradient contribution to the free energy density
W_{h}	Isotropic hardening contribution to the free energy density
W_{p}	Plastic contribution to the free energy density
W_0	Initial defect energy density
$\boldsymbol{x}$	Cartesian coordinates

Greek letters

α	Slip system index
$\bar{\alpha}$	A constant in the defect energy density
$\boldsymbol{\alpha}$	Dislocation density tensor
β	Hardening stress
$\tilde{\gamma}_{\alpha}$	Accumulated plastic slip on system α
$\bar{\gamma}$	Average plastic shear
γ_{d}	Discrete plastic slip
γ_{ac}	Accumulated plastic slip

$\gamma_{r=R}$	Maximum plastic shear under torsion loading
$\dot{\gamma}_\alpha$	Plastic shear rate of system α
γ_α	Plastic slip of system α
$\dot{\gamma}_0$	Reference plastic shear rate
$\tilde{\delta}$	Angle between lines of intersection at grain boundary
$\delta\dot{\zeta}, \delta\dot{\boldsymbol{u}}$	Virtual rates
$\Delta\tau$	Difference between resolved and critical shear stress
$\Delta\tau^{\mathrm{C}}$	Critical resolved shear stress difference
$\Delta\bar{\tau}^{\mathrm{C}}$	Average difference between resolved and critical shear stress
ϵ	Numerical regularization constant for the defect energy
ε	Nominal strain
ε^{p}	Scalar overall plastic strain
$\hat{\varepsilon}, \hat{\lambda}$	Matrices
$\boldsymbol{\varepsilon}$	Infinitesimal strain tensor
$\boldsymbol{\varepsilon}^{\mathrm{e}}$	Elastic part of infinitesimal strain tensor
$\boldsymbol{\varepsilon}^{\mathrm{p}}$	Plastic part of infinitesimal strain tensor
$\hat{\zeta}_{\alpha\beta}$	Geometrical slip system transmission matrix (Ashmawi and Zikry, 2002)
ζ	Micromorphic variable
$\vartheta_1, \vartheta_2, \vartheta_3$	Euler orientation-angles
Γ	Grain boundary plane
Γ_{act}	Part of grain boundary with plastically active integration points
$\tilde{\Theta}$	Initial hardening modulus
Θ	Initial hardening modulus
κ	Angle between direction vectors of two slip systems
$\kappa_{\mathrm{c}}, \omega_{\mathrm{c}}$	Critical angles between two vectors

λ_α	Plastic slip parameter of system α
$\tilde{\lambda}$	Weighted sum criterion (Werner and Prantl, 1990)
ν	Poisson's ratio
$\boldsymbol{\xi}$	Vectorial gradient stress
$\tilde{\boldsymbol{\xi}}$	Numerically regularized, vectorial gradient stress
Ξ	Microtraction
Ξ_0^{C}	Initial grain boundary yield strength
$\Xi_{\mathrm{D}}^{\mathrm{C}}$	Dissipative contribution to the grain boundary yield strength
Ξ_Γ	Grain boundary microtraction
Ξ_Γ^{e}	Energetic grain boundary microtraction
Ξ_Γ^{d}	Dissipative grain boundary microtraction
Π	Mathematical constant
π	Generalized, scalar stress, work-conjugate to the micromorphic variable
ρ	Dislocation density
ρ_{GND}	Dislocation density of geometrically necessary dislocations
ρ_{SSD}	Dislocation density of statistically stored dislocations
σ	Nominal stress
τ, τ_α	Resolved shear stress
$\bar{\tau}$	Average resolved shear stress
τ^{C}	Critical resolved shear stress
τ_0^{C}	Initial critical resolved shear stress
τ_∞^{C}	Saturation stress
τ_α^{d}	Dissipative shear stress of system α
τ^{D}	Drag stress
$\tau_{r=R}$	Maximum shear stress under torsion loading
$\tau^{\mathrm{te}}, \tau^{\mathrm{to}}_{r=R,0.2}$	Tensile / torsion shear (0.2–proof) stress

φ	Misorientation angle
$\varphi_\alpha^{\mathrm{AB}}$	Minimum angle between two slip directions (Ekh et al., 2011)
$\varphi_1, \varphi_2, \varphi_3, \varphi_4, \varphi_5,$	Orientation parameters of a grain boundary (rotations only)
φ_{mis}	Minimum misorientation angle between two crystal orientations
ω	Angle between normal vectors of two slip systems
$\hat{\chi}_{\alpha\beta}$	Geometrical slip system transmission matrix (Mayeur et al., 2015)

Operators

$\lvert\nabla \cdot\rvert_x$	Absolute value of gradient of $(\cdot)$ in x-direction
$\lvert \cdot \rvert$	Absolute value of $(\cdot)$
$[\![\cdot]\!]$	Jump of $(\cdot)$ across an interface
$\Delta(\cdot)$	Algorithmic tangent of $(\cdot)$ / difference between two quantities
$\langle\cdot\rangle$	Macaulay brackets: ramp function of $(\cdot)$
$\dot{(\cdot)}$	Rate of $(\cdot)$
$\boldsymbol{a} \times \boldsymbol{b}$	Cross product of two vectors $\boldsymbol{a}, \boldsymbol{b}$
$\boldsymbol{A} \cdot \boldsymbol{B}$	Dot product of two tensors $\boldsymbol{A}, \boldsymbol{B}$
$\boldsymbol{A} \otimes \boldsymbol{B}$	Dyadic product of two tensors $\boldsymbol{A}, \boldsymbol{B}$
$\boldsymbol{A}^{-1}$	Inverse of a tensor $\boldsymbol{A}$
$\boldsymbol{A} = \mathbb{C}\,[\boldsymbol{B}]$	Linear mapping of a second-order tensor by a fourth-order tensor
$\boldsymbol{AB}$	Linear mapping of a second-order tensor
$\boldsymbol{A}^{\mathsf{T}}$	Transpose of a tensor $\boldsymbol{A}$
$\cos(\cdot)$	Cosine of $(\cdot)$
$\mathrm{curl}\,(\cdot)$	Rotation of $(\cdot)$

$\operatorname{div}(\cdot)$	Divergence of a tensor $(\cdot)$
$\mathrm{d}(\cdot)/\,\mathrm{d}x$	Total derivative of $(\cdot)$ w.r.t., e.g., x
$\delta(\cdot)$	Variation of $(\cdot)$
$\partial(\cdot)/\partial x$	Partial derivative of $(\cdot)$ w.r.t., e.g., x
$\operatorname{grad}(\cdot)\,,\nabla(\cdot)$	Gradient of $(\cdot)$
$\mathrm{sg}(\cdot)$	Sign of $(\cdot)$
$\sin(\cdot)$	Sine of $(\cdot)$
$\mathrm{sym}(\boldsymbol{A})$	Symmetric part of a second-order tensor $\boldsymbol{A}$
$\tan(\cdot)$	Tangent of $(\cdot)$

Bibliography

Abuzaid, W. Z., Sangid, M. D., Carroll, J. D., Sehitoglu, H., Lambros, J., 2012. Slip transfer and plastic strain accumulation across grain boundaries in Hastelloy X. J. Mech. Phys. Solids 60 (6), 1201–1220.

Acharya, A., Bassani, J., 2000. Lattice incompatibility and a gradient theory of crystal plasticity. J. Mech. Phys. Solids 48 (8), 1565–1595.

Aifantis, E., 1984. On the microstructural origin of certain inelastic models. J. Eng. Mater. Tech. 106 (4), 326–330.

Aifantis, E. C., 1987. The physics of plastic deformation. Int. J. Plasticity 3 (3), 211–247.

Aifantis, K., Senger, J., Weygand, D., Zaiser, M., 2009. Discrete dislocation dynamics simulation and continuum modeling of plastic boundary layers in tricrystal micropillars. In: IOP Conference Series: Mater. Sci. Eng. Vol. 3, 012025, 1–6. IOP Publishing.

Aifantis, K., Soer, W., De Hosson, J. T. M., Willis, J., 2006. Interfaces within strain gradient plasticity: theory and experiments. Acta Mater. 54 (19), 5077–5085.

Aifantis, K., Willis, J., 2005. The role of interfaces in enhancing the yield strength of composites and polycrystals. J. Mech. Phys. Solids 53 (5), 1047–1070.

Aifantis, K., Willis, J. R., 2004. Interfacial jump conditions in strain-gradient plasticity and relations of Hall–Petch type. Proc. Nat. Congr. Mech. (June 24–26, 2004, Chania/Crete) 7, 372–376.

Anand, L., Gurtin, M. E., Reddy, B. D., 2015. The stored energy of cold work, thermal annealing, and other thermodynamic issues in single crystal plasticity at small length scales. Int. J. Plasticity 64, 1–25.

Anand, L., Kothari, M., 1996. A computational procedure for rate-independent crystal plasticity. J. Mech. Phys. Solids 44 (4), 525–558.

Arsenlis, A., Parks, D. M., Becker, R., Bulatov, V. V., 2004. On the evolution of crystallographic dislocation density in non-homogeneously deforming crystals. J. Mech. Phys. Solids 52 (6), 1213–1246.

Asaro, R. J., 1983. Crystal plasticity. J. Appl. Mech. 50 (4b), 921–934.

Asaro, R. J., Needleman, A., 1985. Overview no. 42: Texture development and strain hardening in rate dependent polycrystals. Acta Metall. 33 (6), 923–953.

Ashby, M., 1970. The deformation of plastically non-homogeneous materials. Phil. Mag. 21 (170), 399–424.

Ashmawi, W., Zikry, M., 2002. Prediction of grain-boundary interfacial mechanisms in polycrystalline materials. J. Eng. Mater. Tech. 124 (1), 88–96.

Bachmann, F., Hielscher, R., Schaeben, H., 2010. Texture analysis with mtex–free and open source software toolbox. Solid State Phenom. 160, 63–68.

Bachurin, D., Weygand, D., Gumbsch, P., 2010. Dislocation–grain boundary interaction in <111> textured thin metal films. Acta Mater. 58 (16), 5232–5241.

Bamford, T., Clark, W., Wagoner, R., 1988. A thermodynamic model of slip propagation. Scripta Metall. 22 (12), 1911–1916.

Bardella, L., 2010. Size effects in phenomenological strain gradient plasticity constitutively involving the plastic spin. Int. J. Eng. Sci. 48 (5), 550–568.

Bardella, L., Panteghini, A., 2015. Modelling the torsion of thin metal wires by distortion gradient plasticity. J. Mech. Phys. Solids 78, 467–492.

Bardella, L., Segurado, J., Panteghini, A., Llorca, J., 2013. Latent hardening size effect in small-scale plasticity. Model. Simul. Mater. Sci. Eng. 21 (5), 055009, 1–28.

Bartels, A., Bartel, T., Canadija, M., Mosler, J., 2015. On the thermomechanical coupling in dissipative materials: A variational approach for generalized standard materials. J. Mech. Phys. Solids 82, 218–234.

Bayerschen, E., Böhlke, T., 2016. Power-law defect energy in a single-crystal gradient plasticity framework: a computational study. Comput. Mech. 58, 13–27.

Bayerschen, E., McBride, A., Reddy, B., Böhlke, T., 2016a. Review on slip transmission criteria in experiments and crystal plasticity models. J. Mater. Sci. 51 (5), 2243–2258.

Bayerschen, E., Prahs, S., Wulfinghoff, S Ziemann, M., Gruber, P., Walter, M., Böhlke, 2016b. Modeling contrary size effects of tensile- and torsion-loaded oligocrystalline gold microwires. J. Mater. Sci. 51, 7451–7470.

Bayerschen, E., Stricker, M., Weygand, D., Böhlke, T., 2016c. Non-quadratic defect energy: A comparison of gradient plasticity simulations to discrete dislocation dynamics results. Proc. Appl. Math. Mech. 16 (1), 301–302.

Bayerschen, E., Stricker, M., Wulfinghoff, S., Weygand, D., Böhlke, T., 2015. Equivalent plastic strain gradient plasticity with grain boundary hardening and comparison to discrete dislocation dynamics. Proc. Roy. Soc. A 471 (2184), 1–19.

Begau, C., Sutmann, G., Hartmaier, A., 2015. Free energy function of dislocation densities by large scale atomistic simulation. arXiv preprint arXiv:1512.02845 .

Berdichevsky, V., 2016. Energy of dislocation networks. Int. J. Eng. Sci. 103, 35–44.

Beyerlein, I., Mara, N., Wang, J., Carpenter, J., Zheng, S., Han, W., Zhang, R., Kang, K., Nizolek, T., Pollock, T., 2012. Structure–property–functionality of bimetal interfaces. JOM 64 (10), 1192–1207.

Beyerlein, I. J., Alexander, D. J., Tomé, C. N., 2007. Plastic anisotropy in aluminum and copper pre-strained by equal channel angular extrusion. J. Mater. Sci. 42 (5), 1733–1750.

Bieler, T., Eisenlohr, P., Roters, F., Kumar, D., Mason, D., Crimp, M., Raabe, D., 2009. The role of heterogeneous deformation on damage nucleation at grain boundaries in single phase metals. Int. J. Plasticity 25 (9), 1655–1683.

Bieler, T., Eisenlohr, P., Zhang, C., Phukan, H., Crimp, M., 2014. Grain boundaries and interfaces in slip transfer. Curr. Opin. Solid St. M. 18 (4), 212–226.

Bilby, B., Bullough, R., de Grinberg, D. K., 1964. General theory of surface dislocations. Discuss. Faraday Soc. 38, 61–68.

Bitzek, E., Brandl, C., Weygand, D., Derlet, P., Van Swygenhoven, H., 2009. Atomistic simulation of a dislocation shear loop interacting with grain boundaries in nanocrystalline aluminium. Modell. Simul. Mater. Sci. Eng. 17 (5), 055008, 1–8.

Bitzek, E., Gumbsch, P., 2005. Dynamic aspects of dislocation motion: atomistic simulations. Mater. Sci. Eng. A 400, 40–44.

Bollmann, W., 1956. Interference effects in the electron microscopy of thin crystal foils. Phys. Rev. 103 (5), 1588–1589.

Bouaziz, O., 2012. Revisited storage and dynamic recovery of dislocation density evolution law: Toward a generalized kocks–mecking model of strain-hardening. Adv. Eng. Mater. 14 (9), 759–761.

Bouvier, S., Alves, J., Oliveira, M., Menezes, L., 2005. Modelling of anisotropic work-hardening behaviour of metallic materials subjected to strain-path changes. Comput. Mater. Sci. 32 (3), 301–315.

Brandl, C., Bitzek, E., Derlet, P., Van Swygenhoven, H., 2007. Slip transfer through a general high angle grain boundary in nanocrystalline aluminum. Appl. Phys. Lett. 91 (11), 111914, 1–3.

Bridier, F., Villechaise, P., Mendez, J., 2005. Analysis of the different slip systems activated by tension in a α/β titanium alloy in relation with local crystallographic orientation. Acta Mater. 53 (3), 555–567.

Brown Jr, W. F., 1941. The effect of dislocations on magnetization near saturation. Phys. Rev. 60 (2), 139–147.

Bunge, H.-J., 2013. Texture analysis in materials science: mathematical methods. Elsevier.

Burgers, J. M., 1939. Some considerations on the fields of stress connected with dislocations in a regular crystal lattice. I. Koninklijke Nederlandse Akademie van Wetenschappen.

Cabus, C., Réglé, H., Bacroix, B., 2014. The influence of grain morphology on texture measured after phase transformation in multiphase steels. J. Mater. Sci. 49 (16), 5646–5657.

Cermelli, P., Gurtin, M. E., 2002. Geometrically necessary dislocations in viscoplastic single crystals and bicrystals undergoing small deformations. Int. J. Solids Struct. 39, 6281–6309.

Chaboche, J.-L., 1989. Constitutive equations for cyclic plasticity and cyclic viscoplasticity. Int. J. Plasticity 5 (3), 247–302.

Chang, H.-J., Cordero, N. M., Déprés, C., Fivel, M., Forest, S., 2016. Micromorphic crystal plasticity versus discrete dislocation dynamics analysis of multilayer pile-up hardening in a narrow channel. Arch. Appl. Mech. 86 (1), 21–38.

Chen, X., Ngan, A., 2011. Specimen size and grain size effects on tensile strength of Ag microwires. Scripta Mater. 64 (8), 717–720.

Chen, Y., 2013. Deformation Behavior of Thin Metallic Wires under Tensile and Torsional Loadings. Vol. 26. KIT Scientific Publishing, Karlsruhe, Germany.

Chen, Y., Kraft, O., Walter, M., 2015. Size effects in thin coarse-grained gold microwires under tensile and torsional loading. Acta Mater. 87, 78–85.

Clark, W., Smith, D., 1979. Interaction of lattice dislocations with periodic grain boundary structures. J. Mater. Sci. 14 (4), 776–788.

Clark, W., Wagoner, R., Shen, Z., Lee, T., Robertson, I., Birnbaum, H., 1992. On the criteria for slip transmission across interfaces in polycrystals. Scripta Metall. Mater. 26 (2), 203–206.

Davis, K., Teghtsoonian, E., Lu, A., 1966. Slip band continuity across grain boundaries in aluminum. Acta Metall. 14 (12), 1677–1684.

De Bresser, J., 1996. Steady state dislocation densities in experimentally deformed calcite materials: single crystals versus polycrystals. J. Geophys. Res.-Sol. EA (1978–2012) 101 (B10), 22189–22201.

De Hosson, J. T., Soer, W. A., Minor, A. M., Shan, Z., Stach, E. A., Asif, S. S., Warren, O. L., 2006. In situ TEM nanoindentation and dislocation-grain boundary interactions: a tribute to David Brandon. J. Mater. Sci. 41 (23), 7704–7719.

Demkowicz, M., Thilly, L., 2011. Structure, shear resistance and interaction with point defects of interfaces in Cu–Nb nanocomposites synthesized by severe plastic deformation. Acta Mater. 59 (20), 7744–7756.

Dewald, M., Curtin, W., 2007a. Multiscale modelling of dislocation/grain-boundary interactions: I. edge dislocations impinging on Σ11 (1 1 3) tilt boundary in Al. Modell. Simul. Mater. Sci. Eng. 15 (1), 193–215.

Dewald, M., Curtin, W., 2007b. Multiscale modelling of dislocation/grain boundary interactions. II. Screw dislocations impinging on tilt boundaries in Al. Philos. Mag. 87 (30), 4615–4641.

Dewald, M., Curtin, W., 2011. Multiscale modeling of dislocation/grain-boundary interactions: III. 60° dislocations impinging on Σ3, Σ9 and Σ11 tilt boundaries in Al. Modell. Simul. Mater. Sci. Eng. 19 (5), 055002, 1–36.

Dingreville, R., Karnesky, R. A., Puel, G., Schmitt, J.-H., 2016. Review of the synergies between computational modeling and experimental characterization of materials across length scales. J. Mater. Sci. 51 (3), 1178–1203.

Dogge, M., Peerlings, R., Geers, M., 2015. Interface modeling in continuum dislocation transport. Mech. Mater. 88, 30–43.

Ekh, M., Bargmann, S., Grymer, M., 2011. Influence of grain boundary conditions on modeling of size-dependence in polycrystals. Acta Mech. 218 (1-2), 103–113.

Eshelby, J., Frank, F., Nabarro, F., 1951. XLI. The equilibrium of linear arrays of dislocations. Lond. Edinb. Dubl. Phil. Mag. 42 (327), 351–364.

Eshelby, J., Read, W., Shockley, W., 1953. Anisotropic elasticity with applications to dislocation theory. Acta Metall. 1 (3), 251–259.

Essmann, U., Rapp, M., Wilkens, M., 1968. Die Versetzungsanordnung in plastisch verformten Kupfer-Vielkristallen. Acta Metall. 16 (10), 1275–1287.

Estrin, Y., Mecking, H., 1984. A unified phenomenological description of work hardening and creep based on one-parameter models. Acta Metall. 32 (1), 57–70.

Ettehad, M., Al-Rub, R. K. A., 2015. On the numerical implementation of the higher-order strain gradient-dependent plasticity theory and its non-classical boundary conditions. Finite Elem. Anal. Des. 93, 50–69.

Eyckens, P., Mulder, H., Gawad, J., Vegter, H., Roose, D., van den Boogaard, T., Van Bael, A., Van Houtte, P., 2015. The prediction of differential hardening behaviour of steels by multi-scale crystal plasticity modelling. Int. J. Plasticity 73, 119–141.

Ezaz, T., Sangid, M. D., Sehitoglu, H., 2011. Energy barriers associated with slip–twin interactions. Philos. Mag. 91 (10), 1464–1488.

Feltham, P., Meakin, J., 1957. On the mechanism of work hardening in face-centred cubic metals, with special reference to polycrystalline copper. Philos. Mag. 2 (13), 105–112.

Fleck, N., Muller, G., Ashby, M., Hutchinson, J., 1994. Strain gradient plasticity: theory and experiment. Acta Metall. Mater. 42 (2), 475–487.

Forest, S., 1998. Modeling slip, kink and shear banding in classical and generalized single crystal plasticity. Acta Mater. 46 (9), 3265–3281.

Forest, S., 2009. Micromorphic approach for gradient elasticity, viscoplasticity, and damage. J. Eng. Mech. 135 (3), 117–131.

Forest, S., 2016. Nonlinear regularization operators as derived from the micromorphic approach to gradient elasticity, viscoplasticity and damage. Proc. Roy. Soc. A 472 (2188), 1–27.

Forest, S., Guéninchault, N., 2013. Inspection of free energy functions in gradient crystal plasticity. Acta Mech. Sinica 29 (6), 763–772.

Forest, S., Sievert, R., 2003. Elastoviscoplastic constitutive frameworks for generalized continua. Acta Mech. 160 (1-2), 71–111.

Forwood, C., Clarebrough, L., 1981. Prismatic glide and slip transfer across a high-angle grain boundary. Philos. Mag. A 44 (1), 31–41.

Franciosi, P., Zaoui, A., 1982. Multislip in fcc crystals a theoretical approach compared with experimental data. Acta Metall. 30 (8), 1627–1637.

Frank, F., Read Jr, W., 1950. Multiplication processes for slow moving dislocations. Phys. Rev. 79 (4), 722–723.

Fredriksson, P., Gudmundson, P., 2005. Size-dependent yield strength of thin films. Int. J. Plasticity 21 (9), 1834–1854.

Geers, M., Cottura, M., Appolaire, B., Busso, E., Forest, S., Villani, A., 2014. Coupled glide-climb diffusion-enhanced crystal plasticity. J. Mech. Phys. Solids 70, 136–153.

Gemperle, A., Zarubova, N., Gemperlova, J., 2005. Reactions of slip dislocations with twin boundary in Fe-Si bicrystals. J. Mater. Sci. 40 (12), 3247–3254.

Gemperlova, J., Polcarova, M., Gemperle, A., Zarubova, N., 2004. Slip transfer across grain boundaries in Fe–Si bicrystals. J. Alloys Compd. 378 (1), 97–101.

Gottschalk, D., McBride, A., Reddy, B., Javili, A., Wriggers, P., Hirschberger, C., 2016. Computational and theoretical aspects of a grain-boundary model that accounts for grain misorientation and grain-boundary orientation. Comput. Mater. Sci. 111, 443–459.

Groma, I., Csikor, F., Zaiser, M., 2003. Spatial correlations and higher-order gradient terms in a continuum description of dislocation dynamics. Acta Mater. 51 (5), 1271–1281.

Guery, A., Hild, F., Latourte, F., Roux, S., 2016. Slip activities in polycrystals determined by coupling dic measurements with crystal plasticity calculations. Int. J. Plasticity 81, 249–266.

Guo, Y., Britton, T., Wilkinson, A., 2014. Slip band–grain boundary interactions in commercial-purity titanium. Acta Mater. 76, 1–12.

Gurtin, M. E., 2000. On the plasticity of single crystals: free energy, microforces, plastic-strain gradients. J. Mech. Phys. Solids 48 (5), 989–1036.

Gurtin, M. E., 2002. A gradient theory of single-crystal viscoplasticity that accounts for geometrically necessary dislocations. J. Mech. Phys. Solids 50 (1), 5–32.

Gurtin, M. E., 2003. On a framework for small-deformation viscoplasticity: free energy, microforces, strain gradients. Int. J. Plasticity 19 (1), 47–90.

Gurtin, M. E., 2008. A theory of grain boundaries that accounts automatically for grain misorientation and grain-boundary orientation. J. Mech. Phys. Solids 56 (2), 640–662.

Gurtin, M. E., Anand, L., Lele, S. P., 2007. Gradient single-crystal plasticity with free energy dependent on dislocation densities. J. Mech. Phys. Solids 55 (9), 1853–1878.

Gurtin, M. E., Needleman, A., 2005. Boundary conditions in small-deformation, single-crystal plasticity that account for the Burgers vector. J. Mech. Phys. Solids 53 (1), 1–31.

Gurtin, M. E., Ohno, N., 2011. A gradient theory of small-deformation, single-crystal plasticity that accounts for GND-induced interactions between slip systems. J. Mech. Phys. Solids 59 (2), 320–343.

Gurtin, M. E., Reddy, B. D., 2014. Gradient single-crystal plasticity within a mises–hill framework based on a new formulation of self-and latent-hardening. J. Mech. Phys. Solids 68, 134–160.

Hall, E., 1951a. The deformation and ageing of mild steel: II characteristics of the lüders deformation. Proc. Phys. Soc. B 64 (9), 742–747.

Hall, E., 1951b. The deformation and ageing of mild steel: III discussion of results. Proc. Phys. Soc. B 64 (9), 747–753.

Han, W., Reddy, D., 2012. Plasticity: mathematical theory and numerical analysis. Vol. 9. Springer Science & Business Media.

Heidenreich, R., 1949. Electron microscope and diffraction study of metal crystal textures by means of thin sections. J. Appl. Phys. 20 (10), 993–1010.

Hielscher, R., Schaeben, H., 2008. A novel pole figure inversion method: specification of the mtex algorithm. J. Appl. Crystallog. 41 (6), 1024–1037.

Hill, R., 1966. Generalized constitutive relations for incremental deformation of metal crystals by multislip. J. Mech. Phys. Solids 14, 95–102.

Hirsch, P., Horne, R., Whelan, M., 1956. Lxviii. direct observations of the arrangement and motion of dislocations in aluminium. Philos. Mag. 1 (7), 677–684.

Hirth, J. P., 1972. The influence of grain boundaries on mechanical properties. Metall. Trans. 3 (12), 3047–3067.

Hirth, J. P., 1985. A brief history of dislocation theory. Metall. Trans. A 16 (12), 2085–2090.

Hirth, J. P., Lothe, J., 1982. Theory of dislocations. 2nd edition, John Wiley & Sons, New York.

Hochrainer, T., Sandfeld, S., Zaiser, M., Gumbsch, P., 2014. Continuum dislocation dynamics: towards a physical theory of crystal plasticity. J. Mech. Phys. Solids 63, 167–178.

Hounkpati, V., Fréour, S., Gloaguen, D., Legrand, V., 2014. Influence of morphologic texture on stress analysis by x-ray and neutron diffraction in single-phase metallic materials. J. Mater. Sci. 49 (20), 7049–7065.

Hull, D., Bacon, D. J., 2011. Introduction to dislocations. Vol. 37. Elsevier.

Hurtado, D. E., Ortiz, M., 2012. Surface effects and the size-dependent hardening and strengthening of nickel micropillars. J. Mech. Phys. Solids 60 (8), 1432–1446.

Hutchinson, J. W., 2000. Plasticity at the micron scale. Int. J. Solids Struct. 37 (1), 225–238.

Jöchen, K., 2011. Personal communication.

Johnston, W. G., 1962. Yield points and delay times in single crystals. J. Appl. Phys. 33 (9), 2716–2730.

Kacher, J., Eftink, B., Cui, B., Robertson, I., 2014. Dislocation interactions with grain boundaries. Curr. Opin. Solid St. M. 18 (4), 227–243.

Kalidindi, S. R., 1998. Incorporation of deformation twinning in crystal plasticity models. J. Mech. Phys. Solids 46 (2), 267–290.

Kametani, R., Kodera, K., Okumura, D., Ohno, N., 2012. Implicit iterative finite element scheme for a strain gradient crystal plasticity model based on self-energy of geometrically necessary dislocations. Comput. Mater. Sci. 53 (1), 53–59.

Kheradmand, N., Knorr, A. F., Marx, M., Deng, Y., 2016. Microscopic incompatibility controlling plastic deformation of bicrystals. Acta Mater. 106, 219–228.

Knorr, A. F., 2014. Definition eines geometrischen Korngrenzenwiderstandes zur Quantifizierung der kristallographischen Fehlpassung benachbarter Kornorientierungen. Shaker, Aachen, Germany.

Kocks, U., 1976. Laws for work-hardening and low-temperature creep. J. Eng. Mater.-T ASME 98 (1), 76–85.

Kocks, U., Brown, T., 1966. Latent hardening in aluminum. Acta Metall. 14 (2), 87–98.

Koehler, J. S., Apr 1952. The nature of work-hardening. Phys. Rev. 86, 52–59.

Koning, M. d., Miller, R., Bulatov, V., Abraham, F. F., 2002. Modelling grain-boundary resistance in intergranular dislocation slip transmission. Philos. Mag. A 82 (13), 2511–2527.

Kooiman, M., Hütter, M., Geers, M., 2015. Microscopically derived free energy of dislocations. J. Mech. Phys. Solids 78, 186–209.

Kooiman, M., Hütter, M., Geers, M., 2016. Free energy of dislocations in a multi-slip geometry. J. Mech. Phys. Solids 88, 267 – 273.

Kovács, I., Feltham, P., 1963. Determination of the work-hardening characteristics of metals at large strains by means of torsion. Phys. Status Solidi B 3 (12), 2379–2382.

Kraft, O., Gruber, P. A., Mönig, R., Weygand, D., 2010. Plasticity in confined dimensions. Annu. Rev. Mater. Sci. 40, 293–317.

Kröner, E., 1959. Allgemeine kontinuumstheorie der versetzungen und eigenspannungen. Arch. Ration. Mech. An. 4 (1), 273–334.

Kumar, B. R., 2010. Influence of crystallographic textures on tensile properties of 316L austenitic stainless steel. J. Mater. Sci. 45 (10), 2598–2605.

Kuroda, M., Tvergaard, V., 2006. Studies of scale dependent crystal viscoplasticity models. J. Mech. Phys. Solids 54 (9), 1789–1810.

Kuroda, M., Tvergaard, V., 2008. On the formulations of higher-order strain gradient crystal plasticity models. J. Mech. Phys. Solids 56 (4), 1591–1608.

Lall, C., Chin, S., Pope, D., 1979. The orientation and temperature dependence of the yield stress of Ni3 (Al, Nb) single crystals. Metall. Trans. A 10 (9), 1323–1332.

Lavrentev, F., 1980. The type of dislocation interaction as the factor determining work hardening. Mater. Sci. Eng. 46 (2), 191–208.

Lavrentev, F., Pokhil, Y. A., 1975. Relation of dislocation density in different slip systems to work hardening parameters for magnesium crystals. Mater. Sci. Eng. 18 (2), 261–270.

Lazar, M., 2010. Dislocations in generalized continuum mechanics. In: Mechanics of Generalized Continua. Springer, pp. 235–244.

Lee, B.-J., Shim, J.-H., Baskes, M., 2003. Semiempirical atomic potentials for the fcc metals Cu, Ag, Au, Ni, Pd, Pt, Al, and Pb based on first and second nearest-neighbor modified embedded atom method. Phys. Rev. B 68 (14), 144112, 1–11.

Lee, T., Robertson, I., Birnbaum, H., 1989a. Anomalous slip in an FCC system. Ultramicroscopy 29 (1), 212–216.

Lee, T., Robertson, I., Birnbaum, H., 1989b. Prediction of slip transfer mechanisms across grain boundaries. Scripta Metall. 23 (5), 799–803.

Lee, T., Robertson, I., Birnbaum, H., 1990. TEM in situ deformation study of the interaction of lattice dislocations with grain boundaries in metals. Philos. Mag. A 62 (1), 131–153.

Li, Z., Hou, C., Huang, M., Ouyang, C., 2009. Strengthening mechanism in micro-polycrystals with penetrable grain boundaries by discrete dislocation dynamics simulation and Hall–Petch effect. Comp. Mater. Sci. 46 (4), 1124–1134.

Liebe, T., Steinmann, P., 2001. Theory and numerics of a thermodynamically consistent framework for geometrically linear gradient plasticity. Int. J. Numer. Meth. Engng. 51 (12), 1437–1467.

Lim, L., Raj, R., 1985a. Continuity of slip screw and mixed crystal dislocations across bicrystals of nickel at 573 K. Acta Metall. 33 (8), 1577–1583.

Lim, L., Raj, R., 1985b. Interaction between lattice and grain boundary dislocations and their role in mechanical properties of interfaces. J. Phys. Colloques 46 (C4), C4–581.

Lim, L., Raj, R., 1985c. The role of residual dislocation arrays in slip induced cavitation, migration and dynamic recrystallization at grain boundaries. Acta Metall. 33 (12), 2205–2214.

Little, M., Krempl, E., Shih, C., 1981. On the time and loading rate dependence of crack-tip fields at room temperature–a viscoplastic analysis of tensile small-scale yielding. Elastic–Plastic Fracture 1, 615–636.

Liu, D., He, Y., Dunstan, D. J., Zhang, B., Gan, Z., Hu, P., Ding, H., 2013. Toward a further understanding of size effects in the torsion of thin metal wires: an experimental and theoretical assessment. Int. J. Plasticity 41, 30–52.

Liu, D., He, Y., Tang, X., Ding, H., Hu, P., Cao, P., 2012. Size effects in the torsion of microscale copper wires: Experiment and analysis. Scripta Mater. 66 (6), 406–409.

Liu, F., Baker, I., Dudley, M., 1995. Dislocation-grain boundary interactions in ice crystals. Philos. Mag. A 71 (1), 15–42.

Livingston, J., Chalmers, B., 1957. Multiple slip in bicrystal deformation. Acta Metall. 5 (6), 322–327.

Lubarda, V. A., 2016. On the recoverable and dissipative parts of higher order stresses in strain gradient plasticity. Int. J. Plasticity 78, 26–43.

Lucadamo, G., Medlin, D., 2002. Dislocation emission at junctions between $\Sigma = 3$ grain boundaries in gold thin films. Acta Mater. 50 (11), 3045–3055.

Luster, J., Morris, M., 1995. Compatibility of deformation in two-phase Ti-Al alloys: Dependence on microstructure and orientation relationships. Metall. Mater. Trans. A 26 (7), 1745–1756.

Ma, A., Roters, F., Raabe, D., 2006. On the consideration of interactions between dislocations and grain boundaries in crystal plasticity finite element modeling–theory, experiments, and simulations. Acta Mater. 54 (8), 2181–2194.

Maire, E., Withers, P., 2014. Quantitative X-ray tomography. Int. Mater. Rev. 59 (1), 1–43.

Maugin, G., 1990. Internal variables and dissipative structures. J. Non-Equilib. Thermodyn. 15 (2), 173–192.

Mayeur, J., Beyerlein, I., Bronkhorst, C., Mourad, H., 2015. Incorporating interface affected zones into crystal plasticity. Int. J. Plasticity 65, 206–225.

McBride, A., Bargmann, S., Reddy, B., 2015. A computational investigation of a model of single-crystal gradient thermoplasticity that accounts for the stored energy of cold work and thermal annealing. Comput. Mech. 55 (4), 755–769.

McBride, A., Gottschalk, D., Reddy, B., Wriggers, P., Javili, A., 2016. Computational and theoretical aspects of a grain-boundary model at finite deformations. Technische Mechanik 36 (1–2), 92–109.

Medlin, D., Carter, C., Angelo, J., Mills, M., 1997. Climb and glide of a/3 <111>dislocations in an aluminium $\Sigma = 3$ boundary. Philos. Mag. A 75 (3), 733–747.

Menzel, A., Steinmann, P., 2000. On the continuum formulation of higher gradient plasticity for single and polycrystals. J. Mech. Phys. Solids 48 (8), 1777–1796.

Mercier, D., Zambaldi, C., Bieler, T. R., 2015. A Matlab toolbox to analyze slip transfer through grain boundaries. In: IOP Conf. Ser. Mater. Sci. Eng. Vol. 82, 012090, 1–4. IOP Publishing.

Mesarovic, S. D., 2010. Plasticity of crystals and interfaces: From discrete dislocations to size-dependent continuum theory. Theoret. Appl. Mech. 37 (4), 289–332.

Miehe, C., 2014. Variational gradient plasticity at finite strains. Part I: Mixed potentials for the evolution and update problems of gradient-extended dissipative solids. Comput. Methods in Appl. Mech. Eng. 268, 677–703.

Miehe, C., Aldakheel, F., Mauthe, S., 2013. Mixed variational principles and robust finite element implementations of gradient plasticity at small strains. Int. J. Numer. Meth. Engng. 94 (11), 1037–1074.

Miehe, C., Mauthe, S., Hildebrand, F., 2014a. Variational gradient plasticity at finite strains. Part III: Local–global updates and regularization techniques in multiplicative plasticity for single crystals. Comput. Methods in Appl. Mech. Eng. 268, 735–762.

Miehe, C., Welschinger, F., Aldakheel, F., 2014b. Variational gradient plasticity at finite strains. Part II: Local–global updates and mixed finite elements for additive plasticity in the logarithmic strain space. Comput. Methods in Appl. Mech. Eng. 268, 704–734.

Mishin, Y., Farkas, D., Mehl, M., Papaconstantopoulos, D., 1998. Interatomic potentials for Al and Ni from experimental data and ab initio calculations. In: MRS Proceedings. Vol. 538, 535–540. Cambridge Univ Press.

Misra, A., Gibala, R., 1999. Slip transfer and dislocation nucleation processes in multiphase ordered Ni-Fe-Al alloys. Metall. Mater. Trans. A 30 (4), 991–1001.

Mughrabi, H., 1978. The cyclic hardening and saturation behaviour of copper single crystals. Mater. Sci. Eng. 33 (2), 207–223.

Needleman, A., Tvergaard, V., 1993. Comparison of crystal plasticity and isotropic hardening predictions for metal-matrix composites. J. Appl. Mech. 60 (1), 70–76.

Nicola, L., Van der Giessen, E., Gurtin, M. E., 2005. Effect of defect energy on strain-gradient predictions of confined single-crystal plasticity. J. Mech. Phys. Solids 53 (6), 1280–1294.

Niordson, C. F., Kysar, J. W., 2014. Computational strain gradient crystal plasticity. J. Mech. Phys. Solids 62, 31–47.

Nix, W. D., Gao, H., 1998. Indentation size effects in crystalline materials: a law for strain gradient plasticity. J. Mech. Phys. Solids 46 (3), 411–425.

Nye, J. F., 1953. Some geometrical relations in dislocated crystals. Acta Metall. 1, 153–162.

Ohno, N., Okumura, D., 2007. Higher-order stress and grain size effects due to self-energy of geometrically necessary dislocations. J. Mech. Phys. Solids 55 (9), 1879–1898.

Ohno, N., Okumura, D., Shibata, T., 2008. Grain-size dependent yield behavior under loading, unloading and reverse loading. Int. J. Mod. Phys. A 22 (31n32), 5937–5942.

Orowan, E., 1934a. Zur Kristallplastizität. I. Z. Phys. 89 (9), 605–613.

Orowan, E., 1934b. Zur Kristallplastizität. II. Z. Phys. 89 (9), 614–633.

Orowan, E., 1934c. Zur Kristallplastizität. III. Z. Phys. 89 (9), 634–659.

Ortiz, M., Repetto, E., 1999. Nonconvex energy minimization and dislocation structures in ductile single crystals. J. Mech. Phys. Solids 47 (2), 397–462.

Özdemir, İ., Yalçinkaya, T., 2014. Modeling of dislocation–grain boundary interactions in a strain gradient crystal plasticity framework. Comput. Mech. 54 (2), 255–268.

Patel, N. R., Bieniek, M., 1979. An analysis of the viscoplastic behavior of metals. Mater. Sci. Eng. 40 (1), 123–134.

Patriarca, L., Abuzaid, W., Sehitoglu, H., Maier, H. J., 2013. Slip transmission in bcc FeCr polycrystal. Mater. Sci. Eng.: A 588, 308–317.

Peirce, D., Asaro, R., Needleman, A., 1982. An analysis of nonuniform and localized deformation in ductile single crystals. Acta Metall. 30 (6), 1087–1119.

Petch, N., 1953. The cleavage strength of polycrystals. J. Iron Steel Inst. 174, 25–28.

Polanyi, M., 1934. Über eine Art Gitterstörung, die einen Kristall plastisch machen könnte. Z. Phys. 89 (9), 660–664.

Pond, R., Smith, D., 1977. On the absorption of dislocations by grain boundaries. Philos. Mag. 36 (2), 353–366.

Pond, R. C., Medlin, D. L., Serra, A., 2006. A study of the accommodation of coherency strain by interfacial defects at a grain boundary in gold. Philos. Mag. 86 (29-31), 4667–4684.

Quek, S. S., Wu, Z., Zhang, Y. W., Srolovitz, D. J., 2014. Polycrystal deformation in a discrete dislocation dynamics framework. Acta Mater. 75, 92–105.

Ramani, E., 2016. Comparison of geometrical grain boundary slip transmission criteria.

Read, W., Shockley, W., 1950. Dislocation models of crystal grain boundaries. Phys. Rev. 78 (3), 275–289.

Reddy, B., 2011a. The role of dissipation and defect energy in variational formulations of problems in strain-gradient plasticity. Part 1: polycrystalline plasticity. Continuum Mech. Therm. 23 (6), 527–549.

Reddy, B., 2011b. The role of dissipation and defect energy in variational formulations of problems in strain-gradient plasticity. Part 2: single-crystal plasticity. Contin. Mech. Thermodyn. 23 (6), 551–572.

Reddy, B., 2013. Variational formulations for single-crystal strain-gradient plasticity at large deformations. GAMM-Mitteilungen 36 (2), 149–160.

Reddy, B. D., Wieners, C., Wohlmuth, B., 2012. Finite element analysis and algorithms for single-crystal strain-gradient plasticity. Int. J. Numer. Meth. Engng. 90 (6), 784–804.

Reuber, C., Eisenlohr, P., Roters, F., Raabe, D., 2014. Dislocation density distribution around an indent in single-crystalline nickel: Comparing nonlocal crystal plasticity finite-element predictions with experiments. Acta Mater. 71, 333–348.

Rieger, F., Böhlke, T., 2015. Microstructure based prediction and homogenization of the strain hardening behavior of dual-phase steel. Arch. Appl. Mech. 85 (9-10), 1439–1458.

Ronay, M., 1979. Yield stress of thin fcc polycrystalline metal films bonded to rigid substrates. Philos. Mag. A 40 (2), 145–160.

Rösler, J., Harders, H., Bäker, M., 2006. Elastisches Verhalten. In: Mechanisches Verhalten der Werkstoffe. Teubner, pp. 31–61.

Roters, F., 2011. Advanced material models for the crystal plasticity finite element method: development of a general CPFEM framework. Tech. rep., Fachgruppe für Materialwissenschaft und Werkstofftechnik.

Sachs, G., Weerts, J., 1930. Zugversuche an Gold-Silberkristallen. Z. Phys. 62 (7-8), 473–493.

Sangid, M. D., Ezaz, T., Sehitoglu, H., 2012. Energetics of residual dislocations associated with slip–twin and slip–GBs interactions. Mater. Sci. Eng.: A 542, 21–30.

Scardia, L., Peerlings, R. H., Peletier, M. A., Geers, M. G., 2014. Mechanics of dislocation pile-ups: A unification of scaling regimes. J. Mech. Phys. Solids 70, 42–61.

Schmid, E., Boas, W., 1935. Kristallplastizität: Mit besonderer Berücksichtigung der Metalle. Springer Berlin Heidelberg.

Seal, J. R., Crimp, M. A., Bieler, T. R., Boehlert, C. J., 2012. Analysis of slip transfer and deformation behavior across the α/β interface in Ti–5Al–2.5 Sn (wt.%) with an equiaxed microstructure. Mater. Sci. Eng. A 552, 61–68.

Seeger, A., Diehl, J., Mader, S., Rebstock, H., 1957. Work-hardening and work-softening of face-centred cubic metal crystals. Philos. Mag. 2 (15), 323–350.

Senger, J., Weygand, D., Gumbsch, P., Kraft, O., 2008. Discrete dislocation simulations of the plasticity of micro-pillars under uniaxial loading. Scripta Mater. 58 (7), 587–590.

Shen, Z., Wagoner, R., Clark, W., 1986. Dislocation pile-up and grain boundary interactions in 304 stainless steel. Scripta Metall. 20 (6), 921–926.

Shen, Z., Wagoner, R., Clark, W., 1988. Dislocation and grain boundary interactions in metals. Acta Metall. 36 (12), 3231–3242.

Shetty, M., 2013. Dislocations and mechanical behaviour of materials. PHI Learning Pvt. Ltd., New Delhi, India.

Shi, J., Zikry, M., 2009. Grain–boundary interactions and orientation effects on crack behavior in polycrystalline aggregates. Int. J. Solids Struct. 46 (21), 3914–3925.

Shi, J., Zikry, M. A., 2011. Modeling of grain boundary transmission, emission, absorption and overall crystalline behavior in $\Sigma1$, $\Sigma3$, and $\Sigma17b$ bicrystals. J. Mater. Res. 26 (14), 1676–1687.

Shizawa, K., Zbib, H., 1999. A thermodynamical theory of gradient elastoplasticity with dislocation density tensor. I: Fundamentals. Int. J. Plasticity 15 (9), 899–938.

Soer, W., Aifantis, K., De Hosson, J. T. M., 2005. Incipient plasticity during nanoindentation at grain boundaries in body-centered cubic metals. Acta Mater. 53 (17), 4665–4676.

Soer, W., De Hosson, J. T. M., 2005. Detection of grain-boundary resistance to slip transfer using nanoindentation. Mater. Lett. 59 (24), 3192–3195.

Spearot, D. E., Sangid, M. D., 2014. Insights on slip transmission at grain boundaries from atomistic simulations. Curr. Opin. Solid St. M. 18 (4), 188–195.

Stricker, M., 2016. Personal communication.

Stricker, M., Gagel, J., Schmitt, S., Schulz, K., Weygand, D., Gumbsch, P., 2016. On slip transmission and grain boundary yielding. Meccanica 51 (2), 271–278.

Sung, J. H., Kim, J. H., Wagoner, R., 2010. A plastic constitutive equation incorporating strain, strain-rate, and temperature. Int. J. Plasticity 26 (12), 1746–1771.

Sylwestrowicz, W., Hall, E., 1951. The deformation and ageing of mild steel. Proc. Phys. Soc. B 64 (6), 495.

Takasugi, T., Izumi, O., Fat-Halla, N., 1978. Activated slip systems during yielding of α-β brass two-phase bicrystals. J. Mater. Sci. 13 (9), 2013–2021.

Taylor, G., 1938. Plastic strain in metals. Twenty-eighth May Lecture, Inst. Metals 62, 307–324.

Taylor, G. I., 1934. The mechanism of plastic deformation of crystals. Part I. theoretical. Proc. Roy. Soc. A 145 (855), 362–387.

Taylor, J. W., 1965. Dislocation dynamics and dynamic yielding. J. Appl. Phys. 36 (10), 3146–3150.

Teodosiu, C., Sidoroff, F., 1976. A theory of finite elastoviscoplasticity of single crystals. Int. J. Eng. Sci. 14 (2), 165–176.

Thompson, A. W., Baskes, M. I., Flanagan, W. F., 1973. The dependence of polycrystal work hardening on grain size. Acta Metall. 21 (7), 1017–1028.

Tian, L., Russell, A., Anderson, I., 2014. A dislocation-based, strain–gradient–plasticity strengthening model for deformation processed metal–metal composites. J. Mater. Sci. 49 (7), 2787–2794.

Tiba, I., Richeton, T., Motz, C., Vehoff, H., Berbenni, S., 2015. Incompatibility stresses at grain boundaries in Ni bicrystalline micropillars analyzed by an anisotropic model and slip activity. Acta Mater. 83, 227–238.

Toda, H., Kamiko, T., Tanabe, Y., Kobayashi, M., Leclere, D., Uesugi, K., Takeuchi, A., Hirayama, K., 2016. Diffraction-amalgamated grain boundary tracking for mapping 3d crystallographic orientation and strain fields during plastic deformation. Acta Mater. 107, 310–324.

Trémolières, R., Lions, J.-L., Glowinski, R., 1981. Numerical analysis of variational inequalities. Vol. 8 of Studies in Mathematics and Its Applications. Elsevier.

van Beers, P., Kouznetsova, V., Geers, M., 2015a. Defect redistribution within a continuum grain boundary plasticity model. J. Mech. Phys. Solids 83, 243–262.

van Beers, P., Kouznetsova, V., Geers, M., 2015b. Grain boundary interfacial plasticity with incorporation of internal structure and energy. Mech. Mater. 90, 69–82.

van Beers, P., Kouznetsova, V., Geers, M., Tschopp, M., McDowell, D., 2015c. A multiscale model of grain boundary structure and energy: From atomistics to a continuum description. Acta Mater. 82, 513–529.

Van Beers, P., McShane, G., Kouznetsova, V., Geers, M., 2013. Grain boundary interface mechanics in strain gradient crystal plasticity. J. Mech. Phys. Solids 61 (12), 2659–2679.

Voce, E., 1948. The relationship between stress and strain for homogeneous deformation. J. Inst. Met. 74, 537–562.

Voce, E., 1955. A practical strain-hardening function. Metallurgia 51 (307), 219–226.

Voyiadjis, G. Z., Faghihi, D., Zhang, Y., 2014. A theory for grain boundaries with strain-gradient plasticity. Int. J. Solids Struct. 51 (10), 1872–1889.

Šiška, F., Weygand, D., Forest, S., Gumbsch, P., 2009. Comparison of mechanical behaviour of thin film simulated by discrete dislocation dynamics and continuum crystal plasticity. Comput. Mater. Sci. 45 (3), 793–799, proceedings of the 17th International Workshop on Computational Mechanics of Materials IWCMM-17.

Wang, J., Hoagland, R., Liu, X., Misra, A., 2011. The influence of interface shear strength on the glide dislocation–interface interactions. Acta Mater. 59 (8), 3164–3173.

Wang, J., Kang, K., Zhang, R., Zheng, S., Beyerlein, I., Mara, N., 2012. Structure and property of interfaces in ARB Cu/Nb laminated composites. JOM 64 (10), 1208–1217.

Werner, E., Prantl, W., 1990. Slip transfer across grain and phase boundaries. Acta Metall. Mater. 38 (3), 533–537.

West, E., Was, G., 2013. Strain incompatibilities and their role in intergranular cracking of irradiated 316L stainless steel. J. Nucl. Mater. 441 (1), 623–632.

Weygand, D., Friedman, L. H., der Giessen, E. V., Needleman, A., 2002. Aspects of boundary-value problem solutions with three-dimensional dislocation dynamics. Model. Simul. Mater. Sci. Eng. 10 (4), 437–468.

Weygand, D., Senger, J., Motz, C., Augustin, W., Heuveline, V., Gumbsch, P., 2009. High performance computing and discrete dislocation dynamics: Plasticity of micrometer sized specimens. In: Nagel, W., Kröner, D., Resch, M. (Eds.), High Performance Computing in Science and Engineering '08. Springer Berlin Heidelberg, pp. 507–523.

Wo, P., Ngan, A., 2004. Investigation of slip transmission behavior across grain boundaries in polycrystalline Ni3Al using nanoindentation. J. Mater. Res. 19 (01), 189–201.

Wolf, D., 1990. Correlation between structure, energy, and ideal cleavage fracture for symmetrical grain boundaries in fcc metals. J. Mater. Res. 5 (08), 1708–1730.

Wulfinghoff, S., 2011. Personal communication.

Wulfinghoff, S., 2014. Numerically Efficient Gradient Crystal Plasticity with a Grain Boundary Yield Criterion and Dislocation-based Work-Hardening, PhD-thesis. Vol. 5. KIT Scientific Publishing, Karlsruhe, Germany.

Wulfinghoff, S., Bayerschen, E., Böhlke, T., 2013. A gradient plasticity grain boundary yield theory. Int. J. Plasticity 51, 33–46.

Wulfinghoff, S., Böhlke, T., 2012. Equivalent plastic strain gradient enhancement of single crystal plasticity: theory and numerics. Proc. Roy. Soc. A 468 (2145), 2682–2703.

Wulfinghoff, S., Böhlke, T., 2013. Equivalent plastic strain gradient crystal plasticity–enhanced power law subroutine. GAMM-Mitteilungen 36 (2), 134–148.

Wulfinghoff, S., Böhlke, T., 2015. Gradient crystal plasticity including dislocation-based work-hardening and dislocation transport. Int. J. Plasticity 69, 152–169.

Wulfinghoff, S., Forest, S., Böhlke, T., 2015. Strain gradient plasticity modeling of the cyclic behavior of laminate microstructures. J. Mech. Phys. Solids 79, 1–20.

Yalcinkaya, T., Brekelmans, W., Geers, M., 2011. Deformation patterning driven by rate dependent non-convex strain gradient plasticity. J. Mech. Phys. Solids 59 (1), 1–17.

Yalçinkaya, T., Brekelmans, W., Geers, M., 2012. Non-convex rate dependent strain gradient crystal plasticity and deformation patterning. Int. J. Solids Struct. 49 (18), 2625–2636.

Yang, B., Motz, C., Rester, M., Dehm, G., 2012. Yield stress influenced by the ratio of wire diameter to grain size–a competition between the effects of specimen microstructure and dimension in micro-sized polycrystalline copper wires. Philos. Mag. 92 (25-27), 3243–3256.

Yao, W., Krill III, C., Albinski, B., Schneider, H.-C., You, J., 2014. Plastic material parameters and plastic anisotropy of tungsten single crystal: a spherical micro-indentation study. J. Mater. Sci. 49 (10), 3705–3715.

Zaafarani, N., Raabe, D., Singh, R., Roters, F., Zaefferer, S., 2006. Three-dimensional investigation of the texture and microstructure below a nanoindent in a cu single crystal using 3D EBSD and crystal plasticity finite element simulations. Acta Mater. 54 (7), 1863–1876.

Zehetbauer, M., Seumer, V., 1993. Cold work hardening in stages IV and V of FCC metals–I. Experiments and interpretation. Acta Metall. Mater. 41 (2), 577–588.

Zghal, S., Coujou, A., Couret, A., 2001. Transmission of the deformation through γ-γ interfaces in a polysynthetically twinned TiAl alloy: I. Ordered domain interfaces (120° rotational). Philos. Mag. A 81 (2), 345–364.

Zghal, S., Couret, A., 2001. Transmission of the deformation through γ-γ interfaces in a polysynthetically twinned TiAl alloy: II. Twin interfaces (180° rotational). Philos. Mag. A 81 (2), 365–382.

Zhang, K., Holmedal, B., Hopperstad, O. S., Dumoulin, S., Gawad, J., Van Bael, A., Van Houtte, P., 2015. Multi-level modelling of mechanical anisotropy of commercial pure aluminium plate: crystal plasticity models, advanced yield functions and parameter identification. Int. J. Plasticity 66, 3–30.

Zhang, X., Aifantis, K., 2015. Interpreting the internal length scale in strain gradient plasticity. Rev. Adv. Mater. Sci. 41 (1), 72–83.

Zhang, X., Aifantis, K. E., Senger, J., Weygand, D., Zaiser, M., 2014. Internal length scale and grain boundary yield strength in gradient models of polycrystal plasticity: How do they relate to the dislocation microstructure? J. Mater. Res. 29 (18), 2116–2128.

Zhao, M., Li, J., Jiang, Q., 2003. Hall–petch relationship in nanometer size range. J. Alloy Compd. 361 (1), 160–164.

Ziemann, M., Chen, Y., Kraft, O., Bayerschen, E., Wulfinghoff, S., Kirchlechner, C., Tamura, N., Böhlke, T., Walter, M., Gruber, P., 2015. Deformation patterns in cross-sections of twisted bamboo-structured Au microwires. Acta Mater. 97, 216–222.

Zikry, M., Kao, M., 1996. Inelastic microstructural failure mechanisms in crystalline materials with high angle grain boundaries. J. Mech. Phys. Solids 44 (11), 1765–1798.

Schriftenreihe Kontinuumsmechanik im Maschinenbau Karlsruher Institut für Technologie (KIT) (ISSN 2192-693X)

Herausgeber: Prof. Dr.-Ing. Thomas Böhlke

Die Bände sind unter www.ksp.kit.edu als PDF frei verfügbar oder als Druckausgabe bestellbar.

Band 1 Felix Fritzen
Microstructural modeling and computational homogenization of the physically linear and nonlinear constitutive behavior of micro-heterogeneous materials. 2011
ISBN 978-3-86644-699-1

Band 2 Rumena Tsotsova
Texturbasierte Modellierung anisotroper Fließpotentiale. 2012
ISBN 978-3-86644-764-6

Band 3 Johannes Wippler
Micromechanical finite element simulations of crack propagation in silicon nitride. 2012
ISBN 978-3-86644-818-6

Band 4 Katja Jöchen
Homogenization of the linear and non-linear mechanical behavior of polycrystals. 2013
ISBN 978-3-86644-971-8

Band 5 Stephan Wulfinghoff
Numerically Efficient Gradient Crystal Plasticity with a Grain Boundary Yield Criterion and Dislocation-based Work-Hardening. 2014
ISBN 978-3-7315-0245-6

Band 6 Viktor Müller
Micromechanical modeling of short-fiber reinforced composites. 2016
ISBN 978-3-7315-0454-2

Schriftenreihe Kontinuumsmechanik im Maschinenbau
Karlsruher Institut für Technologie (KIT)
(ISSN 2192-693X)

Band 7 Florian Rieger
Work-hardening of dual-phase steel. 2016
ISBN 978-3-7315-0513-6

Band 8 Vedran Glavas
Micromechanical Modeling and Simulation of Forming Processes. 2017
ISBN 978-3-7315-0602-7

Band 9 Eric Bayerschen
Single-crystal gradient plasticity with an accumulated plastic slip: Theory and applications. 2017
ISBN 978-3-7315-0606-5